BAYOU HARVEST

Carl A. Brasseaux and Donald W. Davis, series editors

BAYOU HARVEST

Subsistence Practice in Coastal Louisiana

Helen A. Regis and Shana Walton

University Press of Mississippi / Jackson

This contribution has been supported with funding provided by the Louisiana Sea Grant College Program (LSG) under NOAA Award # NA14OAR4170099. Additional support is from the Louisiana Sea Grant Foundation. The funding support of LSG and NOAA is gratefully acknowledged, along with the matching support by LSU. Logo created by Louisiana Sea Grant College Program.

The University Press of Mississippi is the scholarly publishing agency of the Mississippi Institutions of Higher Learning: Alcorn State University, Delta State University, Jackson State University, Mississippi State University, Mississippi University for Women, Mississippi Valley State University, University of Mississippi, and University of Southern Mississippi.

www.upress.state.ms.us

The University Press of Mississippi is a member of the Association of University Presses.

Copyright © 2024 by University Press of Mississippi
All rights reserved

∞

Library of Congress Cataloging-in-Publication Data

Names: Regis, Helen A., 1965– author. | Walton, Shana, 1961– author.
Title: Bayou harvest : subsistence practice in coastal Louisiana / Helen A. Regis, Shana Walton.
Other titles: America's third coast.
Description: Jackson : University Press of Mississippi, 2024. | Series: America's third coast | Includes bibliographical references and index.
Identifiers: LCCN 2023032833 (print) | LCCN 2023032834 (ebook) | ISBN 9781496849069 (hardback) | ISBN 9781496849076 (trade paperback) | ISBN 9781496849083 (epub) | ISBN 9781496849090 (epub) | ISBN 9781496849106 (pdf) | ISBN 9781496849113 (pdf)
Subjects: LCSH: Food—Social aspects—Louisiana—History. | Food habits—Louisiana—History. | Identity (Psychology)—Social aspects—Louisiana. | Louisiana—Social life and customs.
Classification: LCC GT2850 .R44 2024 (print) | LCC GT2850 (ebook) | DDC 394.1/209763—dc23/eng/20230928
LC record available at https://lccn.loc.gov/2023032833
LC ebook record available at https://lccn.loc.gov/2023032834

British Library Cataloging-in-Publication Data available

CONTENTS

Oyster Spaghetti: A Preface vii

1. Framing Subsistence: "It's Just What We Do" 3

2. Portraits of Practice 19

3. Harvesting as History 33

4. Heritage, Identity, and Place 51

5. Family, Community, and Feasts 63

6. Camps, Leases, and Clubs 79

7. "Worth It" and Other Measures of Value 99

8. Self-Reliance, Care, and Mutual Aid 135

9. Conclusion 153

Postscript: Hurricane Ida 163

Acknowledgments 165

Appendix A: Discussion Questions, Resources, and Project Ideas 167

Appendix B: Eight Factors Used in Customary and Traditional Determinations in Alaska 173

Notes 175

References 183

Index 197

OYSTER SPAGHETTI
A Preface

Research team meeting, Thibodaux, Louisiana, December 11, 2011. Wendy Billiot is telling us about her phone calls with participants who agreed to track their subsistence harvesting, sharing, and consumption:

When I call these people, they don't want to just say, "We went to town and ate at IHOP today." They want to say, "The pancakes were so good, and we had praline syrup." And when they eat something that was all from the grocery store, they want to tell me everything they cooked and where they bought it. They don't realize that's not the [subsistence] project. So, I listen to their stories. [*Wendy reads out loud from her notes.*] Here's one from the *Duprés*:

"For supper, we had boiled shrimp that our son brought to us already boiled. He owns a machine shop and sometimes he does work for friends and they give him seafood. And also for dinner John cooked an oyster spaghetti made from the oysters our son gave us which he got in trade for some other machine work. And we are waiting for people to come pick up satsumas from our trees. We had garlic bread that our grandkids sell as a fundraiser for their school. They sell cookie dough too, but John is a diabetic, so we don't buy the cookies. We buy the garlic cheese bread instead. Even though I love it, I really think it's just a rip off!"

The team breaks up laughing.

Helen: [*Laughing*] Sometimes, down the road, you wind up thinking, wait a minute, they were teaching me something important, but at the time I didn't understand why it was important.

Our research team had been working on food logs. We wanted them to be easy to use and precise enough to systematically record activities. Wendy Billiot, a community scholar, had begun working with her network of neighbors and friends to distribute and collect completed forms. In this meeting, Wendy reported that people were just not filling out the forms. She had to shift strategies and was now making phone calls most days to get a verbal report that she herself would enter on the forms. "We had initially thought that we would put the forms in people's hands," she reflected. "And that immediately was like an 'ugh!' And so, I don't mind the phone calls." We agreed we wanted to record stories like this one, so we would understand not just what people are harvesting, but also what it means to them.

✦ ✦ ✦

The oyster spaghetti story is, in part, a story of fieldwork evolution, common to most ethnographic projects. We quickly learned that no matter how simple the form we created, most community members would only check off the boxes for a few days. We also learned that many people would happily spend time on the phone, nearly every day, telling you stories about what they harvested, hunted, cooked, or shared. That day, our discussion rushed back to what we thought was our real topic, but not without Helen noting a fieldwork truism—you never know until much later what will emerge as important.

In our work, we came back to that story again and again. In part because that oyster spaghetti supper was a window into the incredibly layered nature of what anthropologists call *subsistence*. The rest of this preface is a background of how our project came to be and a close reading of the story itself to show how we came to rely on such storytelling to develop our understanding of hunting and harvesting in Louisiana.

We (Helen and Shana) are your improbable guides to navigating these waters. Neither of us started this project as subsistence experts, food scholars, or rural specialists. Quite the opposite. Both of us live in New Orleans, Louisiana's most urban location. For many years before working on this project, we collaborated in documenting festival culture in the city. We had approached the social science office of the Environmental Studies Program of the Bureau of Ocean Energy Management (BOEM) and proposed they help fund a study to look at the ties between Louisiana festivals and the petroleum industry. Helen got a call from Harry Luton, an anthropologist working out of the agency's Gulf of Mexico office, and he said that our proposal was creative, well developed, and, frankly, not fundable. Then he said, "But I have this other project idea that I think *will* get funded. Do you have a pen?" And he began to talk about *subsistence*—a cultural practice in which humans grow, hunt, catch, and share their food—something we all teach about in Anthropology 101 but rarely as an integral part of twenty-first-century North American culture. Of course, we knew something about hunting, fishing, and gardening. Living in Louisiana, we had fished, knew other folks who fished regularly and some who hunted, and we both gardened. Almost every day, as we drove through New Orleans, we would see people fishing in the city's Bayou St. John. When we interviewed a man in Mid City about festival parking, he opened up the ice chest in the bed of his truck to show us the wild boar he had just killed in New Orleans East. So many people had an uncle or a cousin who shrimped. We took it for granted. There's an old saying that fish never realize that they are swimming in water. Our project was about making that water visible. We proposed a wide-ranging look at what kinds of methods would be useful to document wild harvesting, from production to exchange and consumption, including family feasts, community gatherings, commercial connections, and cultural meanings. For three years, from 2011 to 2014, we organized and led a team of researchers working in two Louisiana coastal parishes located two hours southwest of New Orleans, Terrebonne and Lafourche.

Fast forward about a year to the story at the top of this chapter, which happened at one of our earliest team meetings. We were sitting in the dean's conference room at Nicholls State University in Thibodaux, Louisiana. In addition to Helen and Wendy, the other researchers present were Annemarie Galeucia (then an LSU graduate student), Victor Hernandez (then a Nicholls student), and Shana Walton (Nicholls professor and project codirector). As we noted in the vignette, we were in the process of debriefing about some of the research methods we were testing out: short interviews, transects, community and family-based logs, working with community-based researchers, making inventories, participant observation, focus groups, oral history, and community-member essays and writing. The story came about as Wendy explained what was happening in her community with the food logs (figure 1). People wanted to talk about hunting and fishing, and what they ate, but they looked at the logs and groaned (which we transcribed as "ugh!"). So Wendy had taken to calling and asking about what people had eaten during the week and recording their stories. We all found the story funny, but we quickly moved on. At the time, we were focused on our methods, our research process, and generating data. There is power in numbers, and we were committed to doing rigorous research that would produce numbers—calories consumed, percentages of foods harvested and eaten, rates of participation in subsistence and sharing—so our study could speak to policy makers. This book includes a lot of the information we collected from logs and systematic observations. But what we know as anthropologists and folklorists is that there is power in stories. And now we want to stop and pick our way through the story more carefully in order to really hear what Mrs. Dupré is telling us.

Hunting, Fishing, and Harvesting Log

Name: My family - W. Billiot

What day did you go? (circle)	MON	TUE (circled) Dec. 27th	WED	THU	FRI	SAT	SUN	Where did you go? A friend's hunting lease in Theriot, LA
What did you do with what you got? (circle)	I froze it (circled)		I ate it.		I shared it.		Other (use space below to explain)	Who did you go with? My friend & his dad
Please use this space to write anything else you think is important: We (3 of us) shot our limits of duck. 6 each.								How do you have access to this site? His dad took us. We used my mud boat.

Please Circle (or write in) what you...

Hunted		Fished		Harvested	
Deer	Snake	Shrimp	Catfish	Tomatoes	Peppers
Duck (circled)	Rabbit	Redfish	Crabs	Okra	Beans
Hog	Frog	Trout	Crawfish	Corn	Onions
Raccoon	Gator	Oysters		Sugar Cane	Melons
Squirrel					

Figure 1. Daily food log. This is one version of the log used by participants, designed to be easy to check off items and fill in blanks. This log was trial tested by researcher Wendy Wilson Billiot in documenting a duck-hunting trip with a friend.

The story from the Dupré family is one of the routes we can explore to see how complex, leaky, and multichanneled a study of Louisiana subsistence can be. Let's start with Wendy's call to Mrs. Dupré. The background is that Mrs. Dupré agreed to complete a family food log for one month. But she, like many people we talked to, resisted the idea that her meals could be reduced to a list of foods. "When I call these people, they don't want to just say, 'We went to town and ate at IHOP today.' They want to say, 'The pancakes were so good, and we had praline syrup.'" When Wendy calls Mrs. Dupré to get a report about items and types of food, she instead gets stories about quality and aesthetics, taste and values. The oyster spaghetti and the boiled shrimp, both made from items their son got in trade, carry flavors you couldn't find in a restaurant. The IHOP breakfast may not be homegrown but has story value: pancakes taste delicious with praline syrup. Wendy gets stories about relationships, how harvested food circulates, how it serves as an exchange good. Think about the things we learn from the short story Mrs. Dupré told:

> She has a husband, son, and grandchildren;
> Her son works in a machine shop and commonly barters his off-time labor for food, and shares that food with his parents;

Her husband has diabetes[1] and is one of the household cooks;
They grow satsumas,[2] which are ripe and being shared with neighbors;
Frozen, packaged food is used as a fundraiser for local schools (cookie dough; garlic bread); and fundraising food, even if it's tasty, is overpriced.

We get an actual food list (of sorts), but we also get glimpses into occupations, economics, health, education, family structure, social networks, and values, not to mention food aesthetics and taste. We hear Mrs. Dupré as saying something like, "You researchers seem to focus on one aspect, the source of the food. Food is more than calories or activity. I am focused on taste, connection, relationships, my family, and community." She starts her story doing the specific things we asked, naming a meal (supper) and telling us what they had (shrimp), but we also learn how the shrimp was harvested and gifted (from her son). Wendy reads: "For supper we had boiled shrimp that our son brought to us already boiled." And she continues, telling us about another specific meal, oyster spaghetti. Quickly, though, her own categories and priorities tumble forth—the unpicked satsumas, the grandchildren, the taste. If you reread the story, you realize even in her early listing that she has another agenda. In eight short sentences, Mrs. Dupré tells us about seven relationships or connections to local people and places, with some mentioned more than once: her son, her husband, her son's friends, her son's work, her neighbors/friends (who will soon pick up the satsumas), her grandchildren, her grandchildren's school. Mrs. Dupré is challenging traditional ideas of what she does. Her activities don't fall into traditional understandings of *subsistence*, in part because this set of practices is not easily untangled from the rest of her life. This brief story—full of family, friends, the ordinariness of harvesting, connection and sharing, and the quotidian details of health and conflicts ("I really think it's just a rip off!")—wanders through many of the categories of a meaningful life (relationships, values, community, economics, recreation). And this is how we came to see subsistence—as a set of practices, pervasive and deeply integrated not only with economics and foodways, but also with values and aesthetics, winding through almost everyone's life, living comfortably side-by-side with frozen garlic bread and IHOP pancakes.

In this book, we draw on research methods that systematically document what people do (specific harvesting and sharing activities) but also how they talk about what they do (the stories they tell) to gain important insights into the meanings of subsistence. Focusing on the idea that there is power in the stories, we will use a series of narratives to guide you through several channels of this subsistence landscape, moving through diverse terrain. If you look closely at a map of the Louisiana coast, you can see that the Mississippi River has several long, winding channels that break off—as large as the Atchafalaya

River and as small as Bayou Pointe-aux-Chênes.[3] These were once an integral part of the river. Combined, their silt built the land where we now garden and hunt. And more than a thousand miles of man-made canals connect between those distributaries (bayous) and lakes. Together, they create literally thousands of waterways, small and large, where people fish, crab, and shrimp, all working their way eventually to the Gulf.[4] In this book, we use key stories from our fieldwork to explore a topic as complex as the coast itself. This is not a complete map of bayou subsistence practices. We only explore a few channels, those we came to know through our fieldwork, to share our growing understanding of subsistence practices in coastal Louisiana. There are many other channels you could follow. We hope this book inspires you to document and share your own communities' subsistence practices, following along your own familiar channels, and finding new ones.

BAYOU HARVEST

1

FRAMING SUBSISTENCE
"It's Just What We Do"

In order to talk about what's important to the Duprés and other bayou families, we need to reframe how we talk about hunting and harvesting. Historically, the concept of subsistence has been defined in contrast to other forms of production and exchange; for example, in opposition to commercial fisheries, industrial-scale farming, or the global capitalist marketplace. In this sense, subsistence might be seen as a vestige of an earlier economic system, one that has been partially or mostly replaced by the dominant capitalist economy (what some call postindustrial capitalism, or late capitalism). From this perspective, you could say subsistence practices survive as a part of the informal economy, areas outside the formal economy that are relatively invisible to local, state, and federal governments, uncounted and unrecorded, and therefore not subject to taxation. Scholars working in Canada have called this "the shadows" of capitalism (Murton, Bavington, and Dokis 2016). Mostly invisible, or illegible to the state, these activities are often not valued as part of the real economy. And yet, clearly, they are highly valued among participants. But when subsistence is defined only in terms of what it is not, giving one umbrella name for the range of activities is difficult.

Scholars and policy writers have invented a long series of names for these practices, including artisanal food production, artisanal fisheries, vernacular foodways, folk foodways, self-provisioning, foraging, gathering, culturally embedded harvesting, sharing, and customary food gathering.[1] These distinct but overlapping terms all refer to a constellation of activities that produce food and livelihoods, and that build relationships in families, networks, and communities, through methods that differ in some significant way from industrial, high-tech, capital-intensive, or market-based foods. Not all of the terms reference the forms of exchange—including gift, barter, and reciprocity—that produce and sustain specific social relations as well as redistribute food and other resources. These exchanges, and the social relations they create and sustain, have been central to how anthropologists write and think about subsistence.

The word *subsistence* also carries broader, conventional meanings of someone eking out a living, something needed for survival. Collecting aluminum cans, babysitting, or raising vegetables to supplement welfare or social security income is often called a subsistence strategy. Prior to roads and rails, the American frontier was characterized by what historians call subsistence farming. Sharecroppers lived by subsistence. Here the idea is that one is barely getting by. One subsists but without a surplus.

Sometimes people use the word *subsistence* to point to a type of lifestyle. Indigenous people and First Nations are often assumed to have lived a subsistence lifestyle prior to contact with Europeans and they have been linked to the concept of *subsistence*.[2] But the reality is that Indigenous communities were often involved in long-distance trade networks, and some communities generated significant surplus and organized complex feasts long before the arrival of Europeans. Extensive trade routes date back some three thousand years (in the case of Poverty Point)[3] and even long-distance routes were in use at the time of European contact, like the 440-mile overland Natchez Trace that linked the Mississippi River to the Cumberland River and the Tennessee River. Clearly, people for centuries had been doing much more than local self-provisioning, despite the fact that their economies were often misunderstood or misrepresented by European observers, who defined Indigenous people and their food systems in opposition to European societies (as primitive, backward, or wild, see Dawdy 2010; Pottery 2016).

Not all examples we find are New World or aboriginal: Europeans living on the frontiers of empire were involved in the fur trade while also self-provisioning by hunting and fishing.[4] Not all examples are historical: subsistence fishers are found on today's ocean coasts; subsistence agriculture continues, alongside cash crop production and wage labor, in much of Africa and parts of the Americas. Many of these areas are described as having dual economies, one organized to meet household and community needs and another to sell in the marketplace. This idea of a "dual economy" may understate how much the two influence each other and change over time. As Eric Wolf argued in *Europe and the People without History* (1982), all societies have been impacted by (and responded to) globalization for hundreds of years, and no society stands outside of history.

What we have seen in coastal Louisiana brings us to focus on a different definition of subsistence. As we noted in the preface about the oyster spaghetti, we see *subsistence* as a set of common, everyday hunting, harvesting, and sharing practices that are distinctive because they have become integral not only to local economics, but also to ideas of individual and group identity, to family and childrearing, to community bonding, resilience, self-reliance, and food sovereignty, as well as to aesthetics, recreation, and well-being. Most coastal people who deeply value growing, catching, hunting, and sharing foods are immersed in these practices like fish in water. People like the Duprés, who grew up with and are embedded in subsistence harvesting and sharing, do not see it as a distinct system but rather as something woven into their everyday lives. People told us, "It's just what we do."

This idea of meaningful, culturally integrated practices resonates with the ways many others are currently writing about hunting and harvesting. A growing group of scholars are writing about how the set of cultural practices we call *subsistence* is enmeshed with the larger cultural, political, ecological, environmental, and historical landscapes. For example, Indigenous ethnobotanist Enrique Salmón (2012) draws on his own family's food heritage and his research with Indigenous farmers in the southwestern US and northwest Mexico to show how traditional ways of growing food and preparing meals are also a way of taking care of the environment. Robin Wall Kimmerer (2013) writes about how she has learned to bridge her scientific training in botany with what she has learned from her Indigenous relatives, from ceremony, and her own experience working with plants. Indigenous scholars are linking traditional ways of growing and sharing food—framed as food sovereignty—with goals of "restoring cultural knowledge, protecting environments, and regaining health" (Mihesuah and Hoover 2019). Poets, visual artists, activists, and scholars are all finding that paying attention to subsistence harvesting and sharing makes sense at this moment in our history, as we reckon with rapid social and environmental change, pandemics of diabetes, addiction, suicide, violence, economic crisis, and climate shocks. In *Yakama Rising: Indigenous Cultural Revitalization, Activism, and Healing* (2013), Michelle Jacob, who is a member of the Yakama Nation of Washington State, describes how preserving fish in the traditional way, language study, and learning traditional dances are all part of a project of Indigenous healing. COVID-19 brought a renewed interest in gardening and spending time in nature as city dwellers and suburbanites faced shuttered schools, workplaces, gyms, and shopping malls. But the revitalization of traditional foodways is deeply rooted and predates the latest pandemics.

Among the best documented subsistence practices are those of Alaska Native households and communities. It is also the only state in the US where subsistence is codified in law and subsistence communities (mostly Indigenous) are given priority on state and federal lands. The evidence from Alaska demonstrates that subsistence is not just about protein, calories, or food security:

> ...subsistence resources *and the activities associated with the harvest of these resources* provide more than food. Participation in family and community subsistence activities, whether it be clamming, processing fish at a fish camp, or seal hunting with a father or brother, provide the most basic memories and values in an individual's life. These activities define and establish the sense of family and community. (Fiske and Callaway 2020:160)

In fact, we were advised against studying the Alaska case too closely early in our research, to make sure we did not attempt to replicate the approach. However, while the environments and resources are radically different, the social and cultural significance of subsistence resonates between Louisiana and Alaska (see Appendix B). In other parts of the US, there is also a substantial literature on subsistence fishing—northeast coastal communities, Rhode Island, Connecticut, the Potomac and Anacostia (Fiske and Callaway 2020), and the Caribbean (Puerto Rico and St. Thomas). In their own study of fishing in the DC area, Shirley Fiske and Don Callaway (2020) demonstrate that urban fishers often brought their practices with them when they migrated from rural places in the South. "We also found that none of the fishermen we interviewed used the term 'subsistence fishing' and not many even knew what it meant, except for the popular notion that some people don't have enough money and have to turn to fishing in order to get food. But, they loved to talk about the fish they caught, who ate them, and how they prepared them. Especially the latter."[5]

Like us, a growing number of scholars are looking for ways to write about human entanglements with other species and how best to understand our era of rapid environmental change. People living on the bayou also wrestle with these questions, as they face some of the most rapid land loss in the continental US (Barra 2021; Coastal Master Plan 2023). While there is a growing scientific consensus on climate change, in bayou communities, not everyone agrees on environmental issues that have been politicized by industries and politicians.

We recognize that this topic is at the center of vital scholarly debates about how best to think of the intersections of local and global economies and the significance of Indigenous, traditional, or alternative ways of growing, harvesting, and sharing foods. Many—indeed most people—might assume that small or localized practices are meaningless in the face of agribusiness and commodified food chains, that the bigger and dominant systems are the only ones that really matter. This kind of domination is what some scholars call totalizing, a situation in which a single viewpoint, perspective, or analysis blankets every aspect of reality and human experience, declaring every other perspective meaningless. We take a different tack here. Drawing on scholarship in the humanities and social sciences that has tracked the links between culturally rooted food systems, social ties, solidarity, self-reliance, health and well-being, we recognize that there can be multiple meaningful ways of growing and exchanging food that can coexist with dominant systems. In fact, it is crucial that we take these practices seriously as doing things otherwise can provide a vision for the future.[6]

Learning from J. K. Gibson-Graham (2008), Anna Tsing (2015), and the Community Economies Collective (2019), we seek to document economic and social practices that persist and coexist with other forms. As Gibson-Graham[7] have argued, those totalizing perspectives have often justified dismissing other ways of living as incidental, epiphenomenal, trivial, or irrelevant—as insignificant footnotes to global capitalism.[8] But there are reasons to pay attention to these practices. Working as we do in the birdfoot delta of the Mississippi River, we use the metaphor of water. Viewed from a single moment in time and almost any specific position, the Mississippi River is one stream, steady, powerful, mighty, and quite solid. But if we zoom out spatially and historically, we see that the river is, actually, lots of waters—tributary streams, bayous (distributaries), former channels merging into marsh, crevasses (figure 2). Not only a complex force of nature, the river is also a product of human culture, with levees and dams, canals and spillways, miles of bridges and docks, as well as million-dollar sediment diversion projects.[9]

We abandon the search for simplicity. In this study of the cultural practices of hunting and harvesting in coastal Louisiana, we focus primarily on what people are actually doing and how they understand what it all means. Like scholars writing about "community economies" we contribute to conversations on what people actually do. Instead of focusing on "an overarching system . . . which operates via a set of universal laws" we consider how subsistence is part of "a decidedly more contingent assemblage of processes, practices, and actors" (2019:57). These practices are not pure or isolated from other livelihoods. Nor are they the sole province of any one group. Practitioners include Indigenous people and those whose ancestors were settlers, people who emigrated willingly or were displaced by war, persecution, or colonization, and those who were trafficked in the slave trade. They are Native American/American Indian, Cajun, Black/African Americans, white/Euro-Americans of German, Irish, Sicilian, and French descent, Isleños whose ancestors came from the Canary Islands during the colonial period, Vietnamese and Laotian Americans, and recent immigrants from Latin America. Viewed as a practice, subsistence has been shaped by all the people who have come to reside in coastal Louisiana, by the histories and technologies of dispossession (Woods 2017). "Everyone carries

PLATE 22
SHEET 13

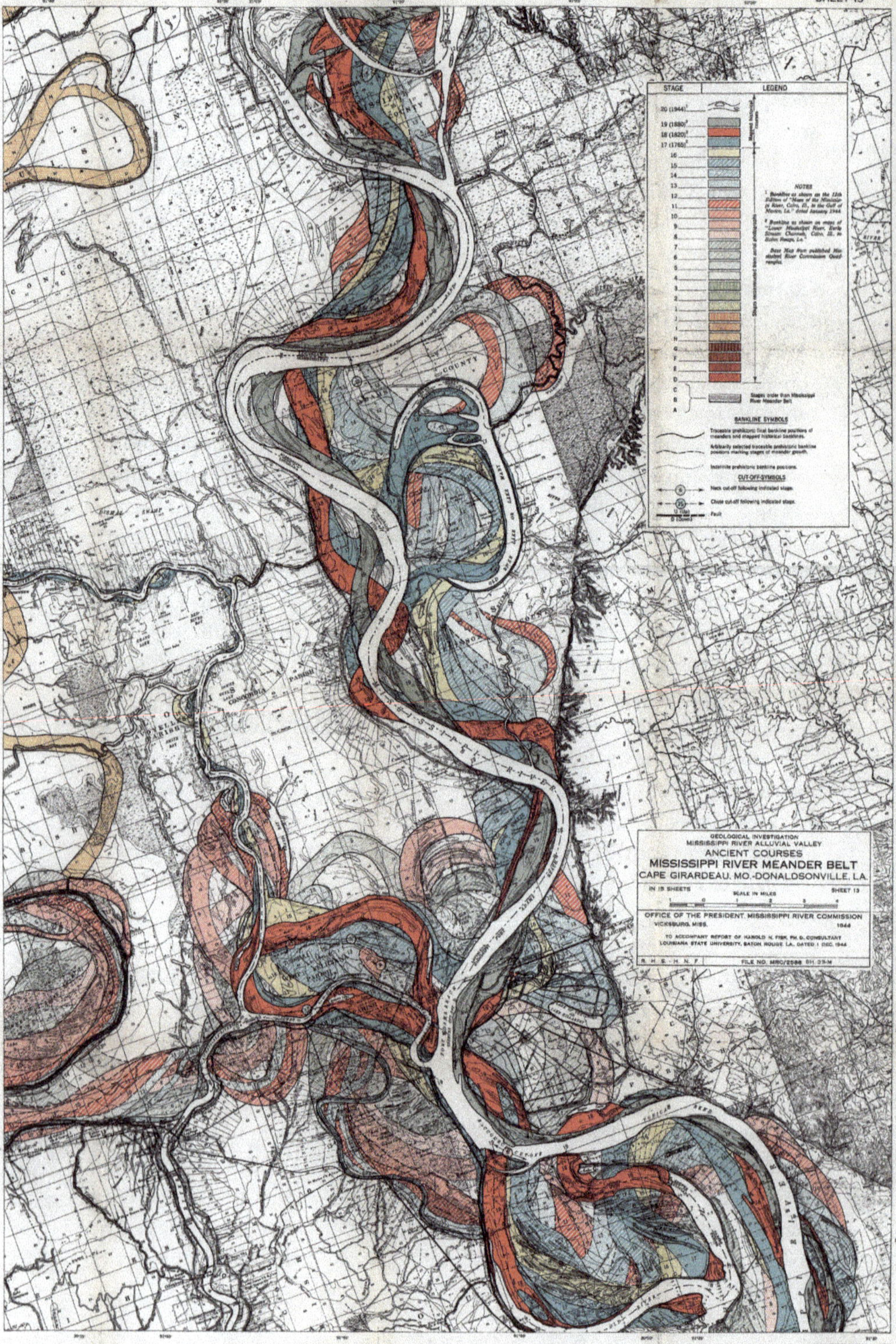

a history of contamination; purity is not an option," writes Anna Tsing, as she explores how mushroom picking strengthens community among Southeast Asian immigrant harvesters (2015:27). This lack of purity, Tsing argues, is what makes survival possible.

Perhaps it is no longer possible to use the term *subsistence* to describe one specific kind of economy or a kind of people (Poe et al. 2013; Robbins et al. 2008). Instead, *subsistence* as we use it describes a diverse array of activities (or practices) governed by nonmarket logics, practices that have goals other than generating profit, and, while contributing to food needs, also contribute to the pleasure of producing fresh, flavorful, highly valued foods and sharing them with friends and family. Those who participate in subsistence activities are themselves embedded in households and social networks that involve multiple livelihoods and mixed economies. In Louisiana, for example, this could mean working the alligator-hunting season to earn enough cash to buy a commercial shrimping license or taking a job offshore, working on deepwater oil and gas, because the pay and intermittent schedule (such as fourteen days on followed by twenty-one days off) generate disposable income for hunting leases and the time to use them. Both of these are actual examples from our study. Household economic strategies are also mixed, and may involve members who shrimp, garden, and work full-time in the wage economy. For many area residents, the significance of crabbing, shrimping, gardening, fin-fishing, or hunting is not only material. Though nearly everyone we talked with eats or shares what they produce, the food represents much more than the calories and the nutrition. It creates and strengthens social ties to family, neighbors, and coworkers. It underwrites extended family gatherings and community feasts. It provides harvesters an opportunity to be outside, to teach children and grandchildren essential skills and values, to have deep experiences with plant and animal life, to simply relax and enjoy, and to connect to their heritage. This heritage is often understood as linked to family, place, ethnicity, and region. It undergirds gender identity, and as we will see, it also connects to notions of autonomy, self-reliance (personal and collective), sovereignty, independence, and solidarity with family and community. And, importantly, subsistence practices contribute to what researchers call well-being and happiness.

Far from a pristine or isolated way of life, subsistence food gathering and harvesting in coastal Louisiana operates in close interaction with

Figure 2. Mississippi River Meander, map created by Harold Fisk (Geological Investigation of the Alluvial Valley of the Mississippi River), US Army Corps of Engineers, 1944. Plate 22, Sheet 13. Image courtesy of the Cartographic Information Center, Department of Geography & Anthropology, Louisiana State University. Public Domain.

commercial fishing, wage-based employment (including the oil industry), and entrepreneurship. Noncommercial household production and consumption of fresh-caught shrimp may often coexist with commercial sales, gift exchange, and barter. There is no bright line between subsistence and other activities or economies. Rather, we argue that any definition of subsistence must rely on a pattern of activities specific to the region. At times, any one of these activities could look like "recreation," but can also coalesce into a pattern of harvesting, use, and sharing that clearly show an alternative system of economics, value, and cultural meaning for people along the coast. Subsistence practices, like markets and identities, are emergent and visible in context. They are also embedded in globally interconnected local histories.

In part, these practices about people and subsistence are reinforced as much by *narrative* as activity and shaped by an economics and ethics of value (Anderson 1995). We can think of both the activities themselves and the ways of talking about the activities as binding together "communities of practice" (Eckert and McConnell-Ginet 1992). Many people across the region—people of diverse ages, ethnicities, genders, socioeconomic classes—build a sense of themselves and others through actions they learn and teach others but also through the ways they talk about those actions and share stories. They have ideas about who is and is not a member of their community. And having a known place in the community, particularly holding authority, is vested with meaning and value. In our study, these communities are both one stream and many. After all, this is not a small homogenous area. Our primary research area covered two parishes that have a combined population of nearly 200,000 people, with networks extending into other parishes along the coast. Many residents are participating in (or identifying with) a wide range of activities, holding specific expertise, authority, and involvement, having motivations ranging from recreation to raising children properly. The channels overlap, diverge, stagnate, spring anew.

Throughout the book, we will continue to visit this broad definition we have outlined for hunting and harvesting practices in coastal Louisiana. At the end of the book, we will return to reflect further on how to think about subsistence. Our ethnography shares what we have learned and how we learned to see subsistence. We consider how the gatherings around fresh-harvested self-provisioned food generate feelings of happiness and well-being. Above all, we show how subsistence practices (and, importantly, talking *about* these practices) are a way for people to be themselves, shaping their identities and values, and binding them to their families and communities. People come to recognize themselves and each other through the stories they tell about hunting, fishing, harvesting, and sharing food. Through harvesting and sharing food, Louisiana residents take care of each other, and in doing so, they become themselves. In this way subsistence practices can be seen as engines of individual and collective self-making.

In the coming chapters, we examine the activities and discourse around these patterns of practice among residents of Terrebonne and Lafourche Parishes as well as other coastal residents (chapter 2); the history of the region in connection with hunting and harvesting (chapter 3); the complicated relationship of heritage, identity, and place in the bayou region (chapter 4); the role of feasting in organizing family, friends, and community gatherings (chapter 5), the central place of fishing and hunting camps in transmitting subsistence heritage (chapter 6), how new evidence generated by our research prompts a rethinking of the economic dimensions of subsistence (chapter 7), and the role of harvesting and sharing in self-reliance, care, and mutual aid (chapter 8). In the conclusion, we consider how our work on subsistence contributes to understanding food sovereignty, the anthropology of happiness and well-being, and research on coastal communities. We end where we began, with the growing sense of urgency that subsistence practice and cultural heritage is at risk from a rapidly changing environment.

Researching Subsistence, Assembling a Team

In the course of this research, our team interviewed dozens of people, recorded more than two thousand log entries, collected hours of taped memories and life histories, and took hundreds of photographs. We accompanied hunts (duck, alligator, rabbit, deer), went crabbing, fishing, and shrimping. We shucked corn, picked tomatoes, attended shrimp boils, and ate everything from venison to thistle. We learned a lot. For example, we can tell you that people harvested or processed fifty-two locally grown vegetables or fruits in a single year. We documented how this abundance fluctuates with the seasons: while people harvest year-round, they reported growing and processing the fewest vegetables in August. Those pieces of information are easy to collect and share.

But as we were gathering information about how often people hunted duck and how many they put in the freezer, we began realizing the breathtaking depth and breadth of what we were trying to describe, its pervasiveness and its near invisibility. In order to see this, we begin with a brief look at the range of experiences, looking for both divergences and connections. In order to talk about those experiences, we have to talk about how we know what we know. In other words, we need to discuss methods. Methods—or the way a project is structured—determine the kind of information that can be collected, the kinds of stories that can be told. We start with telling how we collected information, and then we consider some fieldwork experiences that describe both how people on our team learned and the different kinds of information they learned.

Teamwork Ethnography

Social scientists at the Bureau of Ocean Energy Management had long wanted to do a study of subsistence practices in coastal Louisiana, and the 2010 oil disaster gave these conversations a new urgency. Specifically, researchers at BOEM wanted to lay the groundwork for a coastal-wide study by starting with a smaller area and asking some basic questions: What are people actually doing? What role does this play in their lives? Is it important? And what methods could work to study subsistence?

Because we needed a geographical focus, we proposed Terrebonne and Lafourche Parishes where we already had community and academic contacts. Helen had worked in the coastal region with social scientists from the Bureau for Applied Research in Anthropology (BARA), University of Arizona, and Shana had written her dissertation based on fieldwork in Terrebonne Parish and taught at Nicholls State University in Lafourche Parish (Walton 1994, 2002). We hoped to draw on not only our own research contacts of area leaders and community elders, but those of BARA and the new interdisciplinary Center for Bayou Studies developing at Nicholls, which included faculty from French, biology, and history.

Our proposal and research started with just two people on the project—Helen and Shana—spending weeks talking and gathering ideas. We held countless informal conversations with students and neighbors about their own connections to hunting, fishing, and gardening and to food growing, harvesting, and sharing. Quickly, we knew we needed a team. Learning from research methods that had been employed by BARA to study the socioeconomic impacts of the oil spill, we structured our team around faculty, students, and community members. Diane Austin, a senior scientist at the University of Arizona, and an authority on organizing fieldwork teams, suggested Helen read *Doing Team Ethnography* (Erickson and Stull 1997).

In classic ethnography, anthropologists work alone in the field and solo write most of their books and articles. Despite its prevalence in applied anthropology, the team approach remains novel for many ethnographers. Erickson and Stull provide cautionary advice to researchers on benefits and challenges of working with others, and offer pragmatic advice, such as structures for sharing field notes. As we were developing our project, Helen went to visit a colleague, the renowned Latin Americanist and cultural anthropologist Miles Richardson, who saw the Erickson and Stull book under her arm and said, "There is no such thing as team ethnography."

"I was full of respect and admiration for Miles," Helen remembers, "but I knew that I disagreed with him about this. Some of my best experiences in the field had been working with others—on the Jazz Fest project with Shana and with our students." In fact, Helen's very first ethnographic experience as a graduate student was working on a team-based project to document New Orleans social aid and pleasure clubs, a study funded by the Jean Lafitte National Historical Park and Preserve and the National Park Service.[10]

So, Helen and Shana put together an unusual team for a university-based project. We had two academics from different types of educational institutions, one from a major research university (LSU) and one from a small, regional teaching university (Nicholls State). Our core team included graduate students advised by Helen, a graduate student from another nearby university, two long-established community members, and a new community member who was affiliated with a small subsistence-linked nonprofit.

In order to both recruit participants and to offer full disclosure about the goals of our project, we created a flyer that we posted in libraries and coffee shops, as well as handed out to people on every occasion. The flyer did not use the word "subsistence" at all. All of our informal conversations had already steered us in another direction. Instead, building on work by Maida Owens[11] to explore and help define folklife for communities, the flyer (figure 3) starts out with a series of questions:

Do you . . .
    Harvest or gather your own food?
    Share food with family or church members?
    Keep a freezer full of seafood, vegetables, or game?

The flyer explained:

> We want to hear stories about how such activities fit into people's schedules, what they actually eat, and how they share and exchange among relatives, friends and neighbors. We are interested in exploring how these activities provide not only food, but also community connections and meaningful experiences.

We had other methods of recruitment in addition to the flyer—everything from chance encounters to booths at festivals, hanging out in likely crabbing and fishing spots, recommendations from coworkers, students, and other

Hunted, Harvested and Home Grown:
Food and Community in Coastal Louisiana

This multi-year research project is an effort to better understand the role of harvesting and fishing in the lives of coastal Louisiana residents, and to understand how people use the land and water. Sometimes, non-commercial activities like gardening, catching fish, or giving ducks to a neighbor have been overlooked as important to both the economy and culture of South Louisiana. We want to hear your stories about how these activities fit into your life, what you harvest for your own use, and how you share your harvest with relatives, friends, and neighbors. We are looking at not only what people eat but how food helps to define social ties, well-being, and what it all means.

Our research team is focusing on Terrebonne and Lafourche parishes to document harvesting, hunting and gardening activities. We will draw on the expertise of community residents, voluntary organizations, clubs, business and not-for-profit leaders, and government officials. We welcome the participation of anybody who has a story to tell.

This project is supported by the U.S. Bureau of Ocean Energy Management, Regulation, and Enforcement (BOEMRE, formerly MMS), Nicholls State University and Louisiana State University.

Do you...
▶ harvest or gather your own food?

--fishing?
--hunt for duck, deer, squirrels, or rabbits?
--crabbing, shrimping or frogging?
--gardening?

▶ share food with family or church members?

▶ keep a freezer full of seafood, vegetables or game?

Your story is important

Here are some ways to participate:
(1) Share your story.
(2) Share photographs or other materials that should be copied and archived.
(3) Let us know about others who have stories to tell.

Our team is committed to keeping the identity of all our participants anonymous because many of the stories people share are sensitive. If you would like more information about the project or would like to participate, please contact us:

Shana Walton
shana.walton@nicholls.edu
985-448-4458

Helen Regis
hregis1@lsu.edu
225-819-6357

Figure 3. Recruitment flyer.

fieldworkers in the region, and our own community contacts from years of earlier fieldwork in the region. Because this was a pilot study, we did not systematically recruit people from, say, specific social or demographic groups. Although we wound up with a wide range of people participating, we are keenly aware that not everyone who participates in subsistence down the bayou is represented and future studies will need to take this into account. We explore these issues further in chapter 4.

In general, we had no problems recruiting people. Many more people would have liked to participate than we had capacity to document. Our methods included focus groups, meetings with community organizations, public presentations of the project, walking around dozens of yards and gardens, and long talks in hallways with people about their hunting dogs and hunting camps. We set up tables and booths at community festivals (where people literally lined up to talk). We spent afternoons riding by popular fishing areas doing drop-in interviews with folks fishing by the side of the road and at marinas. We did windshield surveys of fishing, crabbing, and roadside vending. Our community researchers worked with their neighbors to record thousands of log entries documenting hunting, fishing, eating, and food sharing activities. We participated in gardening, hunting, and fishing (from crabbing to rabbit hunting, from fishing and shrimping to alligator hunting, gardening, and ritual feasts). Sometimes we just took notes and photos. Other times, we recorded interviews to transcribe later. We recorded longer, more formal oral histories. We also tried innovative methods such as recruiting whole classes of Nicholls students, who were from bayou communities, to learn the basics of fieldwork and write about how their own families and communities were involved in subsistence practices.

Some methods were more useful than others, but only one method was a total failure: freezer inventories. And it was an idea we all thought was good. Most families we were talking to—including those of our community researchers—had at least one and often two freezers. We even met one family with three freezers. Food that is hunted and harvested comes in at irregular intervals. Families depend on preserving the food to eat later. So, based on the prevalence of freezers and based on the life experiences of community researchers and Shana, whose own family had depended on their freezers for weekly meals, we thought a simple way to measure the importance of hunted and harvested food was to do a monthly freezer inventory, measuring what foods came in and out over, say, a year.

But we quickly learned that a good idea on paper sometimes won't work in the real world. The logic seemed sound, but we had not accounted for the intersection of hunting and harvesting practice with regulations around hunting and harvesting. That is, some people might have a cache of duck meat, deer meat, or fish in their freezer that was over a kill or catch limit. Or maybe they didn't. But it was possible that a record of what went in and what came out of the freezer for several months could implicate people in illegal activity. Even though we were committed to protecting people by keeping their identities private, the very idea of a freezer survey raised flags about surveillance. This coupled with a storehouse of knowledge about the risks of heavy fines for a breach of state Wildlife and Fisheries regulations and a widely shared value of autonomy made this method a no-go. This realization helped us in two ways: first, we quickly jettisoned the freezer count method, which helped us build trust with community members, and second, the feedback about the implications of our plan forced us to raise our awareness of the complex network of regulations that constrain almost all hunting and harvesting activities.[12]

Ultimately, we now think that one of the greatest strengths of our project is not necessarily the wide range of methods, but the diversity of the researchers themselves. We met every few weeks to compare notes, to discuss our experiences, to share what we were learning, and to strategize about how to meet challenges we were encountering in the field. We'd like to introduce you to the members:

Chris Adams holds undergraduate and master's degrees in sustainability, with a focus on community and local food systems. He helped establish and now operates Earthshare Gardens, a small-scale vegetable farm in Lafayette, and through nonprofit organizations he provides support and capacity building to the community of sustainable farmers around south Louisiana. At the time a resident of Thibodaux, Chris was a newcomer to the Lafourche area when he contributed to the study but quickly recognized the key role that farm stands played in the distribution of local produce in the unique landscape of the bayou lands in southeast Louisiana. Currently residing in Lafayette, Chris now works for the city/parish in the long-range planning department on a variety of environmental and sustainability initiatives.

Wendy Wilson Billiot is an author and avid gardener who runs a fishing camp on Bayou Dularge. She has published two children's picture books and multiple articles in *Country Roads Magazine*, *Louisiana Sportsman*, *The Advocate*, and *Houma Times*. At the time of the study, she was a boat captain running educational eco-tours and cohost of "Hunt Fish Talk," a radio program discussing hot topics in the outdoors. Originally from north Louisiana, she relocated to south Louisiana, married a bayou local, a member of the United Houma Nation, and moved to Bayou Dularge, where she reared five children and still resides. She continues her work on her blog "Bayou Woman—Life in the Louisiana Wetlands."

Lora Ann Chaisson is community advocate and a citizen of the United Houma Nation. At the time of the study, she worked as a workforce development specialist with the Department of Labor. She previously ran a domestic violence program working with five tribes in Louisiana. Lora Ann was raised in a down-the-bayou community where fishing, shrimping, gardening, and sharing are an integral part of everyday life and she has traveled widely among Native American communities through her professional work and as a demonstrator and exhibitor of Indigenous basket-making traditions. Lora Ann has served as a longtime member of the Tribal Council, as vice-principal chief, and currently as principal chief of the United Houma Nation.

Jamie Digilormo has roots in the Lake Charles and Shreveport areas and family members who fish and hunt. She learned how to crab from roadside crabbers in southwest Louisiana during her work with the subsistence project. At the time of the study, Jamie was a graduate student in cultural anthropology at LSU, and she is currently working on another BOEM project to code and make digitally accessible years of research about the Gulf Coast.

Tiffany Duet is a native of Lafourche Parish and grew up surrounded by people who hunt, fish, shrimp, and garden. She is an amateur photographer and teaches writing and literature in the Department of English, Modern Languages, and Cultural Studies at Nicholls State University.

Annemarie Galeucia was raised in western Massachusetts, where tensions between working-class communities and affluent residents, including second-homers from New York and Boston, manifested across the cultural landscape. She had a front-row seat watching rural cultures compete with the suburban and urban lifestyles, including debates between generational hunters and antihunting advocates. Annemarie completed her PhD in geography and anthropology at LSU in 2016 and is currently assistant director of Communication Across the Curriculum at Louisiana State University.

Although **Audri Hubbard** is from north Florida, she was not raised in a family that was involved in fishing or shrimping. She joined the subsistence project while pursuing a master's degree in anthropology at LSU and decided to write her thesis on the boat blessings in Louisiana coastal communities, especially in Chauvin (Hubbard 2013). Audri enjoys hiking and exploring the great outdoors with her dogs.

Helen A. Regis teaches at Louisiana State University and writes about collaborative anthropology and public cultures of New Orleans, especially public space, festivals, coastal cities, cultural heritage tourism, belonging, and sustainability. She coauthored *Charitable Choices: Religion, Race, and Poverty in the Post-Welfare Era* and is board chair at the Neighborhood Story Project.

Mike Saunders was new to Louisiana but had grown up hunting and fishing in east Texas and was keenly interested in food practices of Indigenous communities in Guatemala where he was doing dissertation research. He earned his PhD in anthropology from Tulane in 2019 and currently works as program director of the Bayou Culture Collaborative.

Shana Walton did fieldwork on language and culture in Terrebonne Parish and has worked on public folklore and oral history projects in the state, as well as being a coresearcher with Helen on the New Orleans Jazz & Heritage Festival. She is editor of *Ethnic Heritage in Mississippi* and *Languages in Louisiana*. She is a cofounder of the Center for Bayou Studies at Nicholls State University.

In addition to our core members—many of whom worked with us for all three years of the project—we benefited from many contributions of time, contacts, or information (from their own fieldwork or experiences) from numerous faculty members at Tulane University, LSU, and Nicholls State University. One faculty member at Nicholls, Tiffany Duet, shared her network as well as contributing her own interviews and photographs. Shana collaborated with Nicholls State to offer courses focused on writing and documentation about the region.

Throughout this ethnography, we draw on conversations, field notes, and insights generated by this research collective. As some have remarked, both the challenge and the richness of working with others is to maintain the diversity of perspectives that emerge in the field, in notes and logs, and in team meetings, with enough coordination to move the project forward without leveling out the textures and deeper differences that generate new knowledge. Anna Tsing and colleagues have reflected on these issues with their own experience of building a collaborative to study global mushroom commodity chains (Hathaway 2016; Matsutake Worlds 2009). Such approaches have their strengths and weaknesses, but we realized that, for us at least, this approach is both a choice and a natural inclination, part of how we were socialized as scholars. We could say that it's just what we do. We hope this brief introduction to the team will set the stage for understanding the diversity of the topic and to acknowledge both the breadth and the limitations of what we know. Our next chapter is a look at a range of experiences that can be considered *subsistence*.

2

PORTRAITS OF PRACTICE

"Shrimping Was Getting in the Way of His Hunting and Fishing"

Audri Hubbard worked up and down Bayou Little Caillou in 2013 and 2014 documenting the annual Blessing of the Fleet[1] for her master's thesis and collecting stories and logs for our project. In doing the research, we often heard, "you gotta talk to these guys, they're the real deal." Sometimes, we managed to track them down. This portrait of two men from Chauvin, a community known for its shrimpers, shows what they mean: Here are people who participate in multiple community harvesting practices—from gardening to shrimping to alligator hunting. Both are invested in preserving folk traditions, including specific practices such as drying shrimp or the community boucherie.[2] They see these traditions as closely tied to their identity as Cajuns. The following excerpt is from a field report written by Audri based on her interviews and field notes.

It took an entire summer to finally catch up with Jerome Sourdelier. While I waited to meet him, I came to better know his activities through the sharing logs of another person from the Chauvin community. In fact, most often the shrimp, fish, and crab on her dinner table—frequently pre-peeled or fileted—came from Jerome. His catches could also be found re-gifted on those logs as they were prepared into meals and delivered to an elderly woman in the community.

By the time I met Jerome I was already acquainted with the many varieties of hunting and harvesting in which he participates, but I hardly realized the extent to which he harvests his own food. Jerome worked as a commercial shrimper until Hurricane Katrina. After the storm, he sold his commercial trawler, downgraded to a smaller boat, and left the shrimping industry to work on the offshore oil rigs. Despite leaving the seafood industry, Jerome maintains a commercial shrimping license as a backup plan and to legally allow him to catch more than the recreational limit of 100 pounds. Like many people in Chauvin and throughout the region, Jerome had left the shrimping industry full time, although he still trawls each season. Jerome explained that not only is the oil industry more financially secure in this economy but that working a seven-days-on/seven-days-off schedule (common for offshore rigs) leaves much more time for self-provisioning than full-time shrimping. For one thing, full-time shrimpers, who can be out several days at a time, have a hard time keeping up a garden during the spring shrimping season, which usually starts in May.

In addition to shrimping, Jerome hunts duck and deer, fishes speckled trout, redfish, bass, and brim, and grows tomatoes, bell peppers, eggplant, and green onions. He says he *really* likes to hunt, and he and his brother started hunting with his father "when we was old enough to hold up a gun and shoot." Frankly, he thought his full-time shrimping was getting in the way of his hunting and fishing. He keeps three freezers full of game, seafood, and garden vegetables as well as prepared foods, filleted fish, peeled shrimp, and homemade deer sausage. He tries to kill his limit of deer each year (3 does and 3 bucks) and stores it up, doing all of his own processing. He says he never buys beef at the store. As for grocery store trips, Jerome says they purchase few fruits or vegetables, preferring to grow their own. Eggs come from neighbors, and the only meat they buy are occasional pork chops.

During our first meeting he insisted I take home a bag of dried shrimp, a popular traditional food along the Louisiana coast.[3] The practice of sun drying shrimp began as a means of preservation before refrigeration and continued commercially until the FDA ruled that the method violated health codes. Now, shrimp is dried inside shrimp factories according to health guidelines. Although the change occurred years ago, sun-drying shrimp continues in non-commercial ways. Jerome will line his driveway with ant killer and lay out all of the shrimp to sun dry, periodically seasoning the shrimp with salt, pepper, and garlic salt. He described the process of removing the shells to me:

> I put it [the shrimp] in an onion sack and I beat the onion sack against the cement. And as you're beating, the peelings fall apart to almost like a powder or dust. Then just the shrimp stays in the sack. But you know, you can only beat so much out of a sack and then you've got to change the sack, because you'll wear the sack out. I do have a little beater in my shed with some screen on it. It's a hardware cloth they call it. And you turn it where the shrimp just tumbles in there and where it falls apart and breaks up.

In addition to having the dried shrimp as a snack, Jerome said he likes to make an old-fashioned dried shrimp gumbo:

> You know, people my age and older will do it every once or twice a year. It's a different flavor. Everything is different about it when you cook it. You just got in our head, now, once in a blue moon, make a dried shrimp gumbo, you know? Change up the flavor a little bit.

During our interviews Jerome would frequently mention the sharing between friends at small community events, such as weekly dinners and bigger parties. He suggested I speak with Glynn Trahan, a friend of his who participated in many of these social food gatherings. Glynn works commercially in a number of subsistence industries. Currently, his main field is alligator hunting with some commercial shrimping on the side. His family has a long history in Chauvin; therefore, he hunts alligator on his own inherited family land instead of leasing land like many others. Glynn, like Jerome, fishes, hunts, and gardens extensively for his own personal use. He even goes as far as to have his own crawfish ponds for raising crawfish during the Lenten season.[4]

Glynn chose to be an alligator hunter, in part, because he sees that as the only viable commercial harvesting form left. Both of Glynn's parents worked—his mother at the post office and his father in the oil field—and he spent a great deal of time with his grandparents who "lived off the land." They raised their own vegetables, salted their own meat, and even made their own brooms. Glynn associates this with other traits they had, like caring for people in the community. "And I kind of learned that where there was somebody that was in need for something, like the elderly with a broken washing machine or something, I always go help and fix it. That was the things we did down here. Help the people. And I still do it."

Glynn has a lot of cookouts at his house and also participates in several community-wide cooking events. Both he and Jerome described two particular events—bimonthly potluck community dinners and the annual *boucherie*. Glynn elaborated about the potluck gathering:

> At least once or twice a month we'll get a bunch of the friends over and we'll cook. We might have fifteen, twenty, twenty-five people show up here once or twice a month. We'll all get together and we'll cook jambalaya, white beans, fried fish, gumbo, boiled seafood, or whatever. It's just something we do to get together. It's all stuff that we've either grown or caught ourselves. Alligators, ducks, fish, anything that we went out and killed or whatever. We have the meat grinder here. We make our own deer sausage and stuff like that. We grind up deer meat and use it to make spaghetti or breakfast patties in the morning or whatever. We just utilize all the things that we kill and get together and enjoy it as a bunch of friends, you know. Have the camaraderie. Drink a little wine, sometime a little whiskey or beer, things like that. My kids do the same thing. They'll have their friends come over, and they'll play poker of some kind.

Food brings the men together regularly to not only get together but also to share the stories of how they caught their fish or shot their duck. As Jerome pointed out, "bragging rights" are thrown around whenever someone has a particularly good or rare catch. These smaller regular events are complemented by larger gatherings during a boucherie, Mardi Gras, and the Blessing of the Fleet festival.

Both men spoke of the boucherie extensively. Boucheries are a longstanding tradition throughout Cajun history. Textbooks and locals, both, cite the earliest boucheries as a necessity; without refrigeration meat, especially pork, would spoil quickly. The boucherie became a get-together where a pig was slaughtered, roasted, and shared among neighboring families. Now, with refrigerators and freezers, boucheries are no longer necessary, and the event has declined in popularity. Yet, many continue to throw boucheries as a celebration of their Cajunness.

✦ ✦ ✦

Glynn and Jerome are powerful representatives of traditional subsistence culture. As Audri did fieldwork and talked to fishermen and shrimpers, she was told over and over that the people she really needed to talk to were Jerome Sodelier and Glynn Trahan. They were "the real deal," she was told. They still "lived off the land." To many people, Jerome and Glynn's deep participation in everything from gardening to fishing and hunting to commercial shrimping, crawfish farming, and alligator hunting was an ideal they admired. Both men have a wide range of skills developed over years of practice. They have learned to dry their own shrimp and grind their own sausage. They chose occupations that would allow them time to continue their traditions. As Glynn said, he chose alligator hunting specifically because it allowed him to earn a good living while working as a hunter (figures 4 and 5).

But these are not isolated wilderness types. Both Jerome and Glynn learned subsistence practices from family members who linked those practices to the way you train a human to be a caring, responsible community member. Both enjoy deep family ties and community connections through their subsistence practices, and they have a sense of belonging to something larger through shared identity as Cajuns.

While these men are admired for what they do, most people we met in the region were not like Jerome and Glynn. They were more like the members of the TaWaSi social club, an organization that invited us to talk to them about our project, in part because one of our researchers, Tiffany Duet, was a club member. Before Shana started her presentation to the group, at least one member said how good it was that someone was documenting the area's "old traditions." In other words, that member saw the word *subsistence*, which was in the name of the talk, as a historical practice. Clearly, she didn't identify with this. But, by then, our team had learned to reframe "subsistence" because most people were like the TaWaSi member: if we asked, "Do you consider yourself a person who hunts or harvests for part or some of your meals?" then the response was

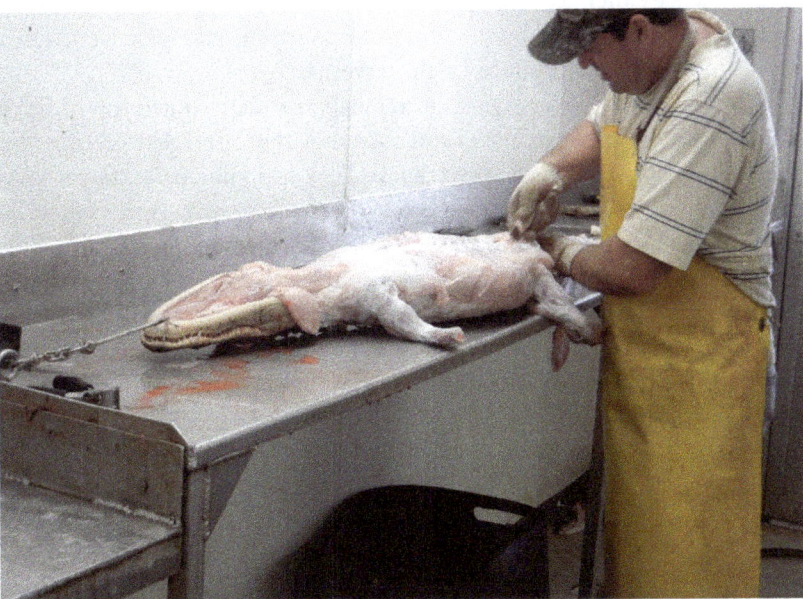

Figure 4 and 5. Skinning an alligator. Like Jerome and Glynn, Joe Autin also shrimps, crabs, and hunts alligator. In these images, he is taking off the hide much like slipping off a coat. A hook holds the alligator steady as he harvests the meat. Alligator hunting is tightly regulated. Photo by Annemarie Galeucia.

generally a clear "No." Of course not, people explained, we mostly get our food at the grocery store. They might add, "We eat a lot of fast food." So, to reframe, Shana asked the members to raise their hands if they

- garden? hunt? shrimp? fish?
- own fishing or hunting equipment you use (or have used)?
- have family who crab for crab boils?
- are sometimes given food that other people have grown or killed?

By the end of this list, every TaWaSi member attending the presentation, about forty-five women, had raised their hands. And this wasn't the only group. For example, in 2012, Shana asked two classes of Nicholls freshmen, fifty students total, "How many of you are involved in (or have family members involved in) hunting or harvesting?" All of them raised their hands. All were full-time students, but they also all knew people or were themselves involved in food production (harvesting, gardening, or hunting). When Helen and team member Audri Hubbard met with a civic group in Dulac in 2011, again, all those in attendance either kept a garden, participated in fishing or hunting, or had friends, family, or neighbors who did and who shared with them a portion of their production (i.e., produce, catch, harvest, or hunt).

Through these stories we get a clearer picture of the channels and eddies in the landscape of subsistence. We learn not only about the range of practices, but that these practices, like bayous and canals in coastal communities, are pervasive and unremarkable. But we also see the power of narrative. For example, the TaWaSi members have a narrative of an idealized practitioner—people like Jerome and Glynn, whose lives are deliberately organized around harvesting and hunting and who see themselves as tradition bearers—the "real deal." In that idealized narrative, speakers often exclude themselves, even though each was connected to harvesting or hunting in some way and regularly ate community-provisioned meals.

Weekend Community

In the fall of 2011, a Nicholls English instructor, Connie Sirois, partnered with our project and tasked her students in English 112, an honors-level composition class, to write a paper based on an interview and observation with a subsistence practitioner. All of the students were from south Louisiana, and most were from either Terrebonne or Lafourche Parish. In the December team meeting, Shana reported on the project:

Connie said what interested her most about the project was that when I walked into the room and she said, "Well, tell them what their topic's going to be for the semester." I said, "It's going to be hunting and fishing, living off the land." And you could hear the audible groan from the students [*laughter*]. Except for one. Rory. He was so excited. He was like, "Yes!" But everybody else just groaned and said, "No, please." And as one student put it when he did his presentation—we had all the students do presentations at the end for the department chair—one of the guys said, "I've spent my whole life just trying not to be part [of this]. My folks all crab and shrimp, and I was always into computers because the one thing I didn't want to do was crab or shrimp. And so, I'm a gamer guy. But you know? I've realized this is important and also really interesting." There was not a single student by the end of the project who did not think it was fascinating.

Much like the TaWaSi members, most of the students had an understanding of the word "subsistence" and either did not see themselves as connected to hunting and harvesting or had distanced themselves from these practices ("because the one thing I didn't want to do was crab or shrimp").

The resulting essays ranged widely from duck hunting and frog gigging to dog training and taxidermy. We were struck by the number of students whose papers focused on the larger context of family and friends, rather than the actions of the practitioner they shadowed. Below is an excerpt, condensed and lightly edited, from an essay written by Rory Eschete—the student who was initially enthusiastic about the project. Rory, who fishes and crabs himself, interviewed his father, who crabbed and fished as a hobby. Interestingly, he did not write about the subsistence practices themselves but centered his essay on weekend gatherings and feastings at a camp.[5]

As I walk up the steps to the deck of my good friend's camp, the distinct smells of boiled seafood, fried fish, and barbecue creep their way to my senses. The anticipation of the atmosphere forces me to speed up my pace because I can't wait to be a part of it. On the end of the deck, all of the men are either telling deep, interesting stories, talking about work, or cracking jokes. Inside the camp, the women are tending to their young children and gossiping about what goes on back in town. All around, everywhere I look, the children are running around or playing games without a worry in the world. I just try to make my rounds and get as much out of the party as possible. I talk to the older men and soak in experience and wisdom from all of the old stories. I watch my dad cook and learn his culinary skill, so I can one day share a similar experience with my family.

Once the food is finally finished, everyone gathers around to share the bountiful harvest. We have fried speckled trout, boiled shrimp, boiled crabs, barbecued deer, fried shrimp, and the list goes on. I make sure I eat till I can't anymore because it isn't every day that people get to have such extravagant celebrations. When my stomach finally alerts me of how full it is, I moved to the couch. On the couch, I begin to dwell upon the idea of how lucky we are to have these occasions. But then, I wander into the questions of why do we have these get-togethers and how did they come to be.

Grand Isle,[6] located off the most southern tip of Lafourche Parish in Louisiana, is known as a warm, welcoming destination for summer family retreats. Many families, like my own, have camps on Grand Isle where we stay on weekends. Down the street from our camp is what I call the largest center of cultural explosion. Rodney, a good friend of my father, has owned this camp for over six years. Located on the Bay side of the island, rather than the beach side, this camp is where I go along with my family and friends to immerse myself in our culture. The camp has everything a Cajun could need. There is a fryer, barbecue pit, smoker, boiling pot, boat, crab traps, and most importantly the twenty-by-forty-foot deck. It's on this deck where people can either get together, socialize, cook, eat, or simply gaze at the stars. Rodney says, "I love to just go fishing with my sons during the day then get back to the camp in the afternoon, drink a few cold sarsaparillas, and begin cooking dinner." The camp isn't anything fancy or extravagant. It offers a place for one to enjoy himself by simply relaxing and having a good time with no stress or worries. After all, that's what everyone goes to the island for. It's a retreat from reality for us to collect our thoughts and focus on life without having the pressures of work, business, and school.

For most families in South Louisiana, our meals include lots of animals and fish native to our area. Much of the responsibility for collecting these prized Cajun delicacies is laid upon the males of the family. In order to have the food supply ready when gatherings come around, the men must either go out into the wild to hunt wild beast and waterfowl or travel by boat in search of fish or frogs. The food brought together at the parties isn't necessarily limited to these items, but they are the usual prospects. For many men, like my father, fishing and crabbing are something they've been doing most of their lives. My father will go down to Grand Isle before the weekend and fish all around the island catching various fish, such as speckled trout, redfish, and flounder. He commented in his interview, "Being able to eat what you catch brings pride and appreciation of [your] work." Usually by the time the weekend comes around and the rest of the family arrives at the camp, there's a promising amount of fish and crabs for everyone to share. The close friends of my dad will bring hunted game, such as deer, frog, or hog in exchange for the fresh seafood being shared. The hard work and passion put into catching and cooking this food makes it taste all the better and everyone appreciates it more. It gives a whole different edge compared to just buying food from the grocery store.

This essay offers a picture of how many people are enmeshed with subsistence while not necessarily being harvesters themselves and how the contributions of a few family members allows large groups of people to participate in activities that depend on harvested food. Most people in the region remain in large community or family networks where at least some members continue subsistence practices recreationally. Unlike the immersive practices of Jerome and Glynn, we found more collaborative, mixed practices, in which participation varied widely depending on a person's age or work. Rory learned to fish and crab, but as a busy college student, he seldom contributes any food. Instead, he shows up on the weekends and eats. That doesn't mean his participation is unimportant or parasitic. In the essay, Rory sees himself as having a role. His job is to pay attention. First, he listens to the stories for "experience and wisdom" and he watches his father cook "so I can one day share a similar experience with my family." In this way, Rory constructs hunting and harvesting as a key part of his identity.

Rory describes the camp at Grand Isle as a "retreat from reality," but the picture we actually see is something functioning more like a traditional Sabbath, a chance to rest, socialize, talk over the last week, celebrate milestones, and reconnect with family and friends. Through this project, we learned not only how pervasive these practices are, we learned how connected they are to wider social worlds. For almost every person we talked to there is little way to separate hunting and harvesting from larger cultural projects, particularly community building, family bonding, recreation, mental health, and enculturation of children.

Asparagus, Bees, and Wildflowers

The previous profile of the camp offers a portrait of Rory, a college student, who is linked to subsistence and defines himself and his family in some ways by their ties to subsistence, but—unlike Jerome or Glynn—is not always an active practitioner. The two profiles share the way in which the participants see subsistence as an essential part of identity. In both, people frame their practices as a type of tradition or heritage and a key to enculturation—teaching a child how to be a member of the local culture. We met other subsistence practitioners, however, who had different motivations and goals. The following profile of Al Guarisco is crafted from field notes by Chris Adams and an oral history recorded by Shana in 2012. Guarisco's practices are not a form of direct inheritance or group identity. Rather here we see approaches that are heavily informed by the circulation of popular ideas of environmentalism and shaped by individual life experiences.

Al Guarisco grew up in coastal Louisiana, around Morgan City, but he explains that his father ran a business. His family did not have time to teach him how to hunt, fish or garden, so he learned from friends and taught himself. He is not carrying on a family tradition, nor did his family use his garden or fishing as a primary food source. Al earned his living as an art teacher and, like many in the region, was a recreational fisherman, hunter, gardener, and in his case, a crawfish farmer, mostly in his spare time. After retirement, he bulldozed the trees on property behind his house in Lafourche Parish and made a large garden, which is still evolving.

The garden is based on not only his memory of gardens growing up, but also the gardens he has researched and those he has seen in Europe with his wife, who is French. He plants traditional crops, but also nontraditional ones, like asparagus. It is springtime, and he is excited about his expanding asparagus beds. He planted two additional beds, with two different seed sources and has about doubled the space he has dedicated to them. Al says he added 1–2 tons of sand to the beds and pine bark, to provide the drainage and loose soil that the asparagus need to not get rotten roots. These are long-term plantings. Asparagus can only be harvested in the second year and will get more productive over several years.

Mostly, the garden is there for Al to eat from and enjoy. But he also uses the garden as supplemental income. For example, Al sells from the roadside in a little stand he made that sits in front of his house. He also sometimes sells at a farmer's market in Houma or Thibodaux, and he sells directly to restaurants. Selling directly to restaurants can be frustrating.

Al has been planting lettuce for several years, often choosing gourmet and Italian head-forming varieties. He told Chris on Oct 30 that he had transplanted 650 of the 1000—that is his target number of seedlings! In the past Al has sold most of his lettuce (although they do eat it at home, just not THAT much), either to restaurants or at markets. He has had trouble with both approaches—at the markets there is not enough or enough consistent demand, so he has a hard time judging how much to harvest and bring. After an unsuccessful market day, he also does not have many options for what to do with the harvested lettuce—it doesn't keep for long and probably will go bad before [the] next opportunity to sell it comes along. On the other hand, some of the restaurants will buy a large quantity one week and none the next, or he simply has a hard time tracking down the chef, or the chef doesn't have enough discretion over purchasing to buy from a local farmer with an irregular production schedule. At times he has received $40–$50 for three ice chests full of lettuce heads. This year he will try to cultivate relationships with two restaurants in Thibodaux and one in Houma (these restaurants range from very fancy to everyday dining). Al is also playing with the idea of selling them from his roadside stand but doesn't think very many people will stop. Because, he says, lettuce is just not something that people are used to buying on the side of the road.

Al's garden looks distinct from other gardens in the region because he follows current ecological gardening practices and interplants wildflowers with the crops, scattering flower seeds in and around the garden and fruit trees. The vegetable, fruit and flower parts of his garden are all mixed, an unusual aesthetic for the region.

Figure 6 and 7. Al Guarisco's asparagus patch and fig tree (*seen in foreground*) are planted in a field of flowers. Photos by Shana Walton.

Beyond vegetables, fruits and flowers, Al also has bees and chickens. And for him, nuisance animals become food. In September 2012, Al was not seeing very many pecans on his trees that year. They should be showing, and he worries there will not be any pecan crop this year. He typically will gather the pecans in the fall, in only a month or so. In years past, he has "battled" with the squirrels over who got to eat them—to the point that he devotes quite a few morning walks around the yard with his pellet gun, aiming to kill the squirrels eating his pecans. He has been fairly successful at times, bagging one or two a week, and eating them as an added bonus. His wife will at times cook them in a stew using a recipe from her home in France that is usually for rabbit. In addition, he sets snares for rabbits that try to sneak under his fence and eat his lettuce. He eats those rabbits too.

First, we note that Al says that none of his family growing up kept gardens, hunted, or fished. He learned some through participating in activities with his friends' families, but he had to seek this out. This information was not passed down to him through his family. Second, he is as motivated by innovation as tradition. He is interested in crops that are not as frequently planted in the region (like several lettuce varieties and asparagus seen in figure 6) and practices not historically common (like beekeeping). His methods and aesthetics (like interplanting with wildflowers and planting trees throughout the gardening area) owe as much to his travels and internet research as they do to traditional models (figure 7). The rabbit prepared in his house is not cooked according to traditional Louisiana foodways but is instead a French dish. Finally, Al's interest is not only feeding family and community but also commerce. Depending on the season, he sometimes has a small roadside stand in front of his house. He sometimes sells in the area farmer's markets and directly to restaurants, sometimes to upscale restaurants with professionally trained chefs seeking to offer customers locally sourced artisanal creations. Al's foodways stand in stark contrast to the grilled and smoked offerings at a community boucherie or a family camp. But still, Al first learned to hunt, fish, and garden from people in the community (even if he uses different techniques now); his sales are small scale; and most of his labor produces food for his own consumption.

We include Al's story to emphasize the pervasiveness of modernity and global connection, a fact almost everyone we talked to hurried to include in their stories. Hunting and harvesting may be a common part of almost everyone's diet and may help shape their calendars, social activities, and gift giving. But Terrebonne and Lafourche Parishes are not isolated, "backward" locations. No one who lives and works in coastal Louisiana is isolated from the world.[7] Project participants sometimes worried that our focus on their hunting and harvesting might imply something retrograde about their region, "Like maybe we don't wear shoes or have cars." As Al Guarisco's story reminds us, people in coastal Louisiana are as much a part of the modern, internet-connected world as anywhere in the US. Teens and college students who hunt also play video games, and they eat as much pizza as venison or fresh shrimp. They go on shopping excursions, take vacations to beaches, mountains, and major cities. Our team members talked not only to people in the woods, on boats, and in gardens. We also met people at restaurants and malls, ate cookies sitting at well-appointed dining-room tables, or pulled up conference chairs in local libraries—everybody with computers and cell phones. People not only save seed, but they also buy seed and bedding plants from garden supply stores and major retailers like Home Depot or Lowe's. A trip through a garden in Chauvin might show you not only local specialties, like cushaw squash and mirliton[8] vines, but also plants from across the US, including a cactus proudly brought back from a trip to the Southwest. Like Al Guarisco, people in Terrebonne and Lafourche look not only to the bayous for inspiration, but also travel to places like France. In fact, our team had to schedule fieldwork with one community member to work around her trips to Hawaii and Europe. Participants went to great pains to make it clear that although these activities link to heritage and culture, neither the practices nor the people are arrested in time.

"What People Do"

In this chapter, we have offered some portraits that show the range of what we will explore as subsistence, the multiple streams that carry meaning, and the discourse people use to explain, frame, and exclude their own activities. We found that few current models or legal definitions of subsistence captured what people actually do or how they think of their practices. We also quickly learned that, when pressed, questions about definitions and distinctions were of little interest to most people. Rather than definitions, they were more keenly interested in documenting their history and making a way for their futures.

3

HARVESTING AS HISTORY

In a public talk on Indigenous People's Day, Houma artist and documentarian Monique Verdin reframed the region's history through the eyes of a foundational historical figure:

> I think of the last matriarchal figure of the Houma, Rosalie Courteau, who in the 1850s, just before the Civil War, bought land—after years of Houma elders trying to petition for these promised land and water rights—in the Yakni Chitto, between the Mississippi River and the Atchafalaya . . . where our people, and many Indigenous people, were pushed to during these times. How she purchased this undesired swamp land at the time. And over time that undesired swamp land becomes a black gold mine. And in the early 1900s. How there's this constant displacement in the history of the Houma, moving from Baton Rouge to [down the bayous]. You know, I have to tell people sometimes:
>
> They're like, "Oh! Houma?! Yeah, I know Houma." I'm like, Yeah. And they're like, "Oh, there's Indians there?" And it's like, What, wait!
>
> They kicked us out of town. And then they renamed the place (Verdin and Breunlin 2020).

Figure 8. Map of Terrebonne and Lafourche Parishes. As is apparent in the satellite image, the traditional boundaries of the parishes extend into an area that is now under water. Houma and Thibodaux, both parish seats, are marked. Lafourche Parish follows the route of Bayou Lafourche. Golden Meadow is marked on lower Bayou Lafourche. Also visible are five of the bayous in Terrebonne Parish. Pointe-aux-Chênes marks the eponymous bayou right at the parish line. Moving west, the bayous are Terrebonne, Little Caillou (marked by Chauvin), then Big Caillou and Dularge. Much of the marsh west of Dularge is uninhabitable. Morgan City marks where the Atchafalaya River flows to the Gulf. The Mississippi is seen flowing southeast out of New Orleans to the birdfoot delta. Map created by Giovanna McClenachan.

Verdin reflects on how this undesired land became a refuge:

Maroons, Cajuns, enslaved Africans, free people of color and Indigenous nations all developed a relationship to place, in the lower Mississippi Delta and among its distributaries and estuaries. The swamp has been the safe refuge, not only Indigenous people like Houma and the Seminole and Maroon communities found everything they needed (Verdin and Breunlin 2020; see Verdin 2019).

So, Who Lives down the Bayou?

In her book, *Return to Yakni Chitto* (2019), Monique Verdin reclaims the Houma name for the area between the Atchafalaya and Mississippi Rivers. Her photography, writing, and public artwork emphasizes the role of the lower Delta as a place of refuge for many nations, and particularly those who were displaced from their homelands through settler colonialism, the transatlantic slave trade, and internal slave markets. Acadians, Vietnamese, and Laotian people have also sought refuge from war, poverty, and political persecution, creating livelihoods in these bayous. German settlers, who were often from the war-torn Alsace region between France and Germany, marked their origins in place names like Des Allemands, but were often assimilated into the region's French/Acadian culture. And more recently Latin American migrants have found work in oil and gas, agriculture, construction, and service industries. But the larger region is better known through a European-centered lens, with a focus on French colonial history, Anglo-American newcomers, and the cultural renaissance of Acadians, better known as Cajuns, whose mid-twentieth-century cultural revitalization placed Cajun cuisine at the center of Louisiana and American foodways and Cajun music in the mix of folk revival. Because New Orleans and Louisiana are frequently featured in American popular culture, people often come with preformed ideas about the people and the land. Many visitors, then, are surprised to learn that New Orleans is not particularly close to the Gulf of Mexico,[1] that there is land south, southeast, and southwest of the city, and that it is not just a vast, uninhabitable swampland.

In this chapter, we consider aspects of the region's history that have shaped contemporary subsistence practices as well as social forces that have led to histories of settlement, displacement, relocation, and dispossession. Those forces have shaped not only who lives there today but also who was displaced and who now travels in to fish and hunt. We realize that identity labels often used by historians may not mesh with how area residents see themselves or with the terms we use in the rest of our account. We use census records, which social scientists and historians consider awkward artifacts of identity and at best a partial lens on a region.

The coastal parishes are, in fact, some of the more heavily populated places in Louisiana, outside of the state's few urban areas (figure 8). Terrebonne Parish, for example, had a population of more than 100,000 people in the 2020 US Census. About a third of those people live in and around the city of Houma. Southern Terrebonne Parish, often called "down the bayou," which was a heavier focus of fieldwork, is home to 2,100 residents.

Lafourche Parish has similar numbers, with a 2020 population of 97,557. Lafourche, which has only one major bayou and sits, more or less, north of Terrebonne, has a slightly less lopsided distribution between the agricultural "up the bayou" and the oil, gas, and seafood "down the bayou" section. About 75 percent of the population lives up the bayou. But due to the linear arrangement of the population on land close to the bayou, the lower part of the parish does not seem particularly rural, earning the nickname of "the longest street in the world" (Davis 2010; Ditto 1980).

Income, Economies, and Health Indicators

Overall, Terrebonne and Lafourche Parishes are, by Louisiana standards, relatively prosperous, ranking in the top 25 percent of median income for parishes in the state. A lot of the region's prosperity is connected to the oil and gas industry with almost 70 percent of all workers in the two parishes working in this sector and related occupations (Gilbert 2006). Although the petrochemical industries dominate the economy, there are also other prominent industries, particularly agriculture and fisheries. For example, Lafourche Parish has more than 27,000 acres cultivating sugarcane, and usually ranks in or near the top ten counties in the US for sugarcane production (US Department of Agriculture 2012). Fisheries remain key for both parishes. In 2010, shrimpers in Terrebonne Parish landed 26.2 million pounds of shrimp worth $28 million, the most shrimp landed in the state, while Lafourche fishers landed $14.8 million in the same year (Oyunginka et al. 2011). Combined, the two parishes account for one-third of all shrimp landed in the state. Shrimping is, of course, not the full extent of the fisheries' value in either parish. In addition, people work in oystering, crabbing, and fishing, particularly charter fishing. In fact, about 20 percent of all seafood harvested in the United States for human consumption comes from coastal Louisiana, with shrimp and blue crabs being the most coveted species (Gramling and Hagelman 2005). Nevertheless, the number of people working in the fisheries industry is declining as well as the wages people are able to earn as independent fishers.

In sum, there are robust economies in both parishes. However, jobs and wealth in the area are not evenly distributed. The poverty rate among the Native American population is a shocking 56.6 percent. Almost 10 percent of households in lower Terrebonne earn below $15,000 per year. Solet (2006:90) notes that "Louisiana's coastal periphery [including southern Terrebonne] has been a large underclass of poor people living on the fringe of capitalism."

Because we are exploring food harvesting practices, it is worth noting that area residents suffer disproportionately from congestive heart failure and diabetes, illnesses that are linked to social inequality, chronic stress, and diet.[2] While in Louisiana overall the number-one cause of death is cancer, in Terrebonne and Lafourche Parishes, congestive heart failure is the leading cause of death. People in our study area suffer from obesity at a higher rate than the state overall, and quite a bit higher than the nation. In Terrebonne Parish, more than 37 percent of the population has a BMI of 30 or greater.[3] Diabetes is a critical problem. The overall rate of diabetes in the US is 10 percent (CDC 2020), and Louisiana outpaces that at 12.3 percent of the population, but in Lafourche Parish that rate climbs to 17 percent. Even more striking is the rate of diabetes in nonwhite populations: 25.9 percent in Lafourche and 18.1 percent in Terrebonne Parish (LA DHH 2014).

The picture that emerges is that the lower parts of the parishes—down the bayou—have fewer residents, significant percentages of white and Native American residents,[4] low numbers of African American residents, and even fewer Latino residents (most of whom are concentrated in small communities); the region struggles with deep poverty, social inequality, and staggering numbers of people (down the bayou and in the parish as a whole) who are suffering from chronic conditions such as obesity and diabetes, metabolic diseases linked to poverty.

To understand how the populations (and wealth) came to be distributed this way and how the economies have changed over time, we briefly explore the region's history.

Unceded Native American Land

The land that now comprises Terrebonne and Lafourche Parishes has been occupied by humans for as long as humans have been in the larger Gulf South region—at least twelve thousand years (Kniffen and Hilliard 1988). However, the actual land that is now thought of as the Gulf Coast has always been a place of flux and change. The land itself consists of delta lobes (and subdelta lobes) created over five thousand years, as the Mississippi River, depositing silt, created distributaries (bayous), which in turn deposited silt.

As those courses became slower because of silt build up, the river would shift course and begin the process anew. Deltas all over the world have long been considered "productive," which means they are good places for humans to find food, animals to hunt, plants to harvest, and rich soil for planting. But they are also vulnerable because the river that grows the land will inevitably shift course (Syvitski et al. 2009), diverting water to other areas and possibly flooding inhabited land. This is the story of Louisiana's eastern coastal parishes, with the land, the bayous, and human habitation shaped by the shifts of the Mississippi.

The area of Terrebonne and Lafourche was formed as the Lafourche lobe, a subdelta region of the larger Mississippi Delta. At the time the Lafourche subdelta was formed, the main channel of the Mississippi ran along what is now called Bayou Lafourche. The river formed multiple distributaries—what we now call bayous, from a Choctaw word *bayuk*. The Lafourche subdelta began forming around AD 400, and soon began gradually shifting course. By AD 1400 the river had completely shifted to the present course. During the one thousand years that the river mainly or even partially flowed through the Lafourche path, the river created streams, lower and higher elevation lands, ponds, swamps, and "landscapes . . . with significant biodiversity and available foodstuffs" (Mehta and Chamberlain 2019:457).

Recent archaeological work shows that between AD 1000 and AD 1400 people lived in settled villages along several of the bayous, including Lafourche, Grand Caillou, Little Caillou, Terrebonne, and Pointe-aux-Chênes. On Bayou Grand Caillou, Indigenous people built a large ceremonial mound, about thirty meters from the banks of the bayou, that would have risen as much as forty feet, visible from quite a distance (Mehta and Chamberlain 2019). Other bayous also have mounds. Such mounds signal that the societies of that time were socially complex. These weren't isolated hunter-gatherer bands. Rather there was a community of people for whom the mound was important. There were leaders to organize work, sufficient workers to act as builders, and enough surplus food to feed people who were hauling dirt to build instead of hunting or gathering food for themselves. While archaeologists haven't found solid evidence of established agriculture along the bayous, there is evidence that people were harvesting abundantly, including hickory nuts, pecans, persimmons, palm and palmetto fruits, acorns, deer, alligator, turtles, catfish, bass, garfish, ducks, herons, and egrets (Dundee et al. 1989; Fritz 1990; Lowery 1974).

In the late 1300s and 1400s, the people in these villages moved and largely abandoned the lower portions of the Lafourche subdelta bayous. Some archaeologists hypothesize that the people moved because as the river changed course, less freshwater flowed through the channels. In other words, the water in their bayous became brackish. The residents had to contend with increasing saltwater intrusion (Mehta and Chamberlain 2019).

By the time European travelers arrived, the lower coastal and birdfoot delta regions now known as lower Plaquemines, lower Lafourche, and lower Terrebonne Parishes were probably inhabited seasonally. People were still using the region as a food resource, using the natural levees to garden and hunt but may not have lived in the lower delta year-round, allowing their communities to shift as the streams and rivers shifted (Usner 2018). Coastal areas have evidence of shell and bone middens at sites where clams, oysters, and turtles were gathered and processed to eat, so the area was well used. Whether or not they lived in the coastal zone year-round, Indigenous people were actively engaged in harvesting from the land and water.

As far as we know, none of the early explorers—La Salle, Iberville, or Bienville—actually navigated down Bayou Lafourche. However, both Iberville and Bienville met Native peoples who lived in the region and they named some of those groups in their reports, including the Washa, Chawasha, Chitimacha, and Bayougoula. So, although some of the large water-based settlements were no longer occupied year-round, large numbers of Native people were living in the Pontchartrain basin and in the areas of Terrebonne and Lafourche (Usner 2018). Davis (2010) estimates that between two hundred and three hundred Washa and Chawasha peoples occupied the area around Bayou Lafourche in the late 1600s and early 1700s. One of the groups who occupy the region now, the Houma, were not in this area at the time of European contact and early colonization. In 1700, the Houma village was on the east side of the confluence of the Red River and the Mississippi River, roughly where the Louisiana State Penitentiary, better known as Angola, is now (Kniffen, Gregory, and Stokes 1987). Another contemporary American Indian tribe, the Chitimacha, had a wider range of territory than their current reservation. Documented Chitimacha settlements in the early 1700s were scattered on the Mississippi, upper Bayou Lafourche, and on Bayou Teche. In fact, some Indigenous activists note that the city of New Orleans was constructed on unceded Chitimacha territory (Bulbancha 2018). From both colonial accounts and archaeology of the city, we know that Native people from around the region used part of the high land where the French Quarter now sits as a marketplace during early colonial times (Usner 2018). While the explorer Bienville was famous for making cooperative agreements with Native people for land use, the larger territory controlled by the Chitimacha, Washa, Chawasha, Bayougoula, Houma, Biloxi, and other groups was never formally ceded to European powers.

Colonization and Land Use

By the early eighteenth century, settlement patterns were beginning to shift rapidly. As more Europeans came in and as warfare, slavery, and disease decimated populations, Indigenous people were on the move. After 1700, many groups shifted. The Washa and Chawasha disappeared as independent groups—some moved to villages along the Mississippi River, then were sold into slavery, and some sought refuge with the Atakapa and Houma. The Houma moved into the Bayou Lafourche and Bayou Terrebonne area, seeking a new home after an attempted massacre by the Tunica. The Tunica themselves had been attacked by the Natchez and relocated, intermarrying with the Biloxi and settling in Avoyelles Parish. Almost every historical tribe in Louisiana shifted location. In fact, only the Chitimacha were able to hold onto any of their ancestral lands (Kniffen et al. 1987). During this time of Indigenous displacement and European in-migration, we see Indigenous people become the first people to use the interior bayou areas as year-round settlements since the Mississippi River's last major shift.

Colonizers long recognized Louisiana as a paradise for fish and game. In his *History of Louisiana*, Le Page du Pratz in 1758 described no fewer than eleven types of fish commonly eaten by people living in the colony (p. 40). Historian Gayarre records in his memoir of 1800, the famous Baron Pontalba reports that "every sort of game and fish is so plentiful that they scarcely fetch any price at all" (Gayarre 1854:434; see also Dawdy 2010). In the early years, however, to the Europeans, all of Louisiana seemed a land more hospitable to hunters and trappers than settlers. Other than a few German immigrants who arrived in the early 1700s, few Europeans came to tackle the wilds of Louisiana. However, from the beginning, the Europeans brought enslaved Africans to work both in the city and in the fledgling plantations forming up and down the Mississippi River. Some of the wilderness areas behind the new plantations and the cypress swamps surrounding New Orleans were considered almost impenetrable and provided shelter for maroons, self-emancipated people, or "runaway" slaves, who began establishing small colonies (Diouf 2014; see also Barnes and Breunlin 2016).

The areas farther south and southwest of New Orleans, including the regions now known as Terrebonne and south Lafourche, remained mostly in the hands of Indigenous people, except for a few trappers. The food supply chains from Native Americans from the Pontchartrain basin and the coastal areas were critical for the survival of the first colonists and, later, the plantations. Usner (2018) notes that Bienville's fledgling New Orleans colony depended on provisions of fresh fish and game as well as harvested corn and other vegetables from tribes ranging from the Acolapissa, the Houma, the Tunica, the Chitimacha, the Biloxi, to the Chaouacha and others. Their knowledge of how to hunt, fish, and grow crops successfully was key to the colony surviving at all.

When the Spanish gained control in 1762, they began more intensive efforts at colonization, including granting land settlements to the Acadians, starting in 1764, and those efforts brought new, intensive settlement into the areas of Terrebonne and Lafourche. Those first Acadian settlements were in the prairie lands, and their homesteads on Bayou Lafourche were high up, not in the marsh or swamplands. In fact, by 1769, only seventeen Acadian families had settled between Donaldsonville and Labadieville (Brasseaux 1985). More began coming to Bayou Lafourche: by 1785, some 600 Acadians were settled between Labadieville and Lafourche Crossing, and in 1794, another 297 Acadians were given land along Lafourche (Brasseaux 1985). In other words, the Acadians moved slowly down Lafourche and moved even more slowly into the interior bayou areas of what is now Terrebonne. This means that well into the middle and late 1700s, the land in lower Terrebonne and Lafourche remained in the hands of Native Americans.

The Louisiana Purchase and the Surge of Sugar Production

By the time of the Louisiana Purchase in 1803, Acadians had become the dominant cultural group along Bayou Lafourche, most living as peasant farmers with small landholdings.[5] For cash crops, they first grew indigo, turning to sugarcane as the process for granulating was improved. In the 1820s, sugar production surged, and most small Acadian landholders along upper Bayou Lafourche were bought out by American and Creole planters, though a few Acadians consolidated landholdings.

At the time of the Louisiana Purchase, the area that was to become Terrebonne Parish had only a few hundred inhabitants. However, most of the land, all the way to the end of the roads, was divided into Spanish land grants that were recognized by the incoming American government. Land that had been lightly inhabited began to experience heavy in-migration, specifically to grow sugarcane. In 1822, Terrebonne Parish was separated from Lafourche Parish. By 1851 the new parish had 110 plantations (Wurzlow 1985) and a population of more than four thousand people (US Census 1850). As the sugar plantation economy took shape stringing plantations and sugar production up and down the bayous, south to Montegut, and farther south to Pointe-aux-Chênes, small landholding Acadians and Native Americans were pushed off land. For example, in the early 1800s, Jean Billiot, a Native American, was recognized as the owner of a large tract of land south of Montegut on Bayou Terrebonne. However, within a few decades, that land became Eloise Plantation and grew sugarcane (Cenac 2017). As Native people and non-landholding Acadians were pushed farther down the bayous and into the marshlands, they began to establish communities such as Pointe-aux-Chênes, Grand Caillou, Dulac, and the island community of Isle de Jean Charles, which at that time had no road connecting to the mainland (Bazet 1934; Brasseaux 1992; Guidry 1980; Kniffen and Hilliard 1988). One early property owner in Pointe-aux-Chênes was Alexandre Billiot, named chief of the Chitimachas in land records. Billiot and his brothers grew sugarcane, corn, and rice and operated a sugar mill, transporting raw sugar to New Orleans to process (Cenac 2017).

As part of the sugar plantation economy, hundreds of enslaved people were brought into the parish. The first enslaved Africans arrived in Louisiana in the early 1700s. But only in the nineteenth century, as the plantation economy expanded, were large numbers of enslaved people living in the Terrebonne-Lafourche area. The Trans-Atlantic Slave Trade had ended in 1808, meaning the people who came to labor on the sugar plantations were purchased through the domestic slave trade, usually through the New Orleans slave markets. Trafficked into the lower Mississippi Delta from other parts of the South, enslaved people in Lafourche-Terrebonne had to orient themselves to new landscapes, crops, and plantation communities (Hall 1992; Jackson 2012). Like the new plantation owners, most of the enslaved workers in Terrebonne and Lafourche were English speakers when they arrived.[6] Across the South, people living in bondage sought their freedom through self-emancipation and the creation of autonomous societies beyond the reach of the plantation. Those who found a way to live in the marshes and swamps were called *maroons*. While there is documented evidence of small and large maroon communities in the cypress swamps up and down the Mississippi River, there is little research on the existence of these communities in Terrebonne-Lafourche.[7] Nevertheless, it seems likely that such communities existed in the region because of their prevalence in swamplands in other parts of the South and throughout the Americas (Bilby 2008; Diouf 2014; Price 1973; Sayers 2016).

During this period, land use in Terrebonne Parish changed dramatically. Land that had primarily served as a seasonal game resource began to be used intensively—for both agriculture and subsistence. While the plantations to the north planted every inch of land in sugarcane, to the south, although there were many plantations, there were also Acadians and Native Americans who lived as farmers and fishers, providing important support services and food supplies to the plantations and to the New Orleans markets (Davis 2010). Some Native families were able to negotiate with planters to live on marshland bordering their plantations in exchange for provisions of fish and game (Dardar 2007; Kniffen et al. 1987).

In addition, enslaved people depended on hunting to provision themselves with food during slavery and in the years following emancipation, as confirmed by slave narratives and archaeological research (Hall 1995; Scott 2008; Wagner 2019; Young et al. 2001). From this, we know that the back areas of coastal plantations may not have been suitable for agriculture but they were well used by Indigenous and African-descendant people, including enslaved and maroon communities, all of whom harvested food both to eat and to sell as well as cypress, cane, and palmetto to make houses and sellable goods. Hall's research (1992) on maroon activity in swamp areas bordering plantations near New Orleans describes both self-provisioning and production for sale in the market, especially through the logging of cypress. In the lower Mississippi Delta, maroon communities surrounded plantations:

> They cultivated corn, squash, and rice, and gathered and ground herbs, roots, and wild fruits for food. They made baskets, sifters, and other articles woven from willow and reeds. They carved indigo vats and troughs from cypress wood. [They] gathered berries, dwarf palmetto roots and sassafras, trapped birds, hunted and fished, and went to New Orleans to trade and to gamble. (Hall 1992:203)

Hall's research shows enslaved people residing on plantations were in frequent contact with maroons living in adjacent wetlands. Colonial records show that these communities were enmeshed through exchange of goods and labor (lumber, firewood, baskets, garden vegetables, cattle, arms, and ammunition) and even "ate together in the woods every day" (Hall 1992:211–22). "They did not distance themselves from the plantations and town; they surrounded them" (Hall 1992:203). This description is likely accurate for maroon and Native American activity in the Terrebonne-Lafourche area as well. Because although upper areas of Terrebonne Parish were transformed by the plantation economy, much of the lower parish was marsh or swamp. Cash crops, such as sugarcane, were planted where there was available land, but much of the lower parish was used for hunting, trapping, fishing, and harvesting cypress lumber.

Along Bayou Lafourche the pattern was much the same, though sugarcane and plantations were to have a much larger role. Well into the 1820s, small farms dotted the upper part of the bayou. In fact, until 1825, cotton, corn, and rice were the primary crops (Ditto 1980). That changed as sugarcane became increasingly dominant in the 1840s. By 1860, sugar was the most important agricultural crop in Lafourche or Terrebonne (Davis 2010) and remains so today (US Department of Agriculture 2012).

In the 1800s, Lafourche and Terrebonne Parishes had a mix of nationalities (Davis 2010). By 1788, Bayou Lafourche had attracted Acadian, Creole French, German, Portuguese, and Spanish migrants, and, by the 1800s, Chinese, Italians, Sicilians, Irish, Yugoslavians, and Americans added to the diversity (Brasseaux 1992). By the end of the 1800s, immigrants from Portugal, Cuba, Denmark, Greece, and Russia were living in wetland enclaves, often in lands that were for all intents and purposes uncharted by Europeans (Davis 2010). Although most German immigrants to Louisiana lived in New Orleans or on the German coast, some made their way to the bayou region (Merrill 2004). For a brief time, enough Germans lived in Houma to support a German-language newspaper (Wurzlow 1984). From 1800 until 1840, New Orleans was one of the most ethnically diverse and largest cities in the US. It was a magnet for immigrants, some of whom relocated to adjacent parishes in the marshes farther south (Campanella 2008; Davis 2010). In the case of Italians, some were directly recruited into the bayou region as labor on sugarcane farms (DeSantis 2016).

Despite this diversity, the French language was dominant, be it Acadian, Plantation French, French Canadian, Creole, or Native American French (Dajko 2019). While the Acadians had faced a forced expulsion from Nova Scotia in 1755 and arrived in Louisiana after a generation of displacement, some French Canadians in the 1800s saw Louisiana as a business opportunity and voluntarily immigrated to buy land and become sugar planters. Other francophones immigrated and made their way to the Lafourche region to recreate their planter lifestyle after the Haitian Revolution (1791–1804; Campanella 2008; Davis 2010; Hunt 1988). Up until the Civil War, most of the communities were Gallicized, meaning that residents of German or Spanish origin would have been assimilated into French-speaking Louisiana culture. This was visible in changes to surnames. For example, the Spanish name Placencias became Plaisance. Rodriguez became Rodrigue. German names also became assimilated. The Louisiana French family name Labranche is a translation from the German name *Zweig*, meaning twig or branch.

Both Houma and Thibodaux, the parish seats, grew as cities during the 1800s. Thibodaux grew large enough to attract steamboat entertainers (Ditto 1980). Towns featured outdoor markets where down-the-bayou residents came to sell game, fish, and vegetables (Wurzlow 1984). With the disruptions of the Civil War, land changed hands and many plantations went bankrupt. However, many of the surviving plantations continued to produce sugarcane, seeing opportunity for growth and expansion. Some families, like the Minors of Terrebonne Parish, were able to hold on to their plantations after the war, and other families, like the Ellenders, were able to expand their sugar plantations and landholdings by purchasing plantations in financial trouble (Cherry 2015).

Hunting for Sport, Regulation, and Conservation

After the Civil War, the South saw the rise of hunting as a form of recreation. Hunting was the single most popular sport of the late 1800s and early twentieth century among southern men (Ownby 1990). In addition, hunting became a form of tourism in the late 1800s and early 1900s (Giltner 2008). In Louisiana this led to the beginnings of out-of-town people establishing hunting camps in coastal marsh towns.

This growth and social change led to the first hunting regulations. During the early years of European settlement of Terrebonne Parish, people hunted game, including wild hogs and wild cattle, pretty much without constraints. The first hunting regulation came in 1824 when the newly formed police jury passed a resolution allowing that free-ranging, unbranded cattle were considered wild game and could be shot on sight. Another antipoaching resolution allowed anyone to hunt wild cattle after taking an oath before a judge that they intended to do so (Cenac et al. 2013). Most hunting regulations in Louisiana, however, came about because of the same two forces that brought on changes in other states in the US South: (1) a desire to regulate and curtail hunting by freedmen—who were often accused of over hunting—and (2) concern over the increase in sport hunting by northern vacationers (Giltner 2008).

This growth in recreation harvesting, along with the rise of the conservation movement in the US, led to Louisiana enacting some of the first hunting and fishing regulations across the South and the creation of state-funded enforcement agencies (Giltner 2008). During the first twenty to thirty years of the twentieth century, Louisiana began to enforce the first limits on hunters; for instance, on how many ducks hunters could kill or what techniques they could employ.

Eddie Henry, who was born in 1924 and raised in Montegut, Terrebonne Parish, remembers duck-hunting strategies from when he was a child that would not be allowed today:

> I remember my daddy used to hunt ducks, and there was no limits on the ducks. In those days you could hunt with live decoys. My dad had some little yard ducks. And they had a little female duck. And what we'd do is he'd go hide her behind the bushes, and the other ducks couldn't see, the other ducks they keep, "Quack, quack, quack," constantly. And of course that brought in the ducks. (Sell and McGuire 2008)

The transition to livelihoods based around trapping and fishing took place during the latter years of the nineteenth century. By the beginning of the twentieth century this would be well established as the traditional Houma lifestyle and continues to be so today. Davis (2010) asserts that all wetland communities at the turn of the century in Louisiana relied on fishing, trapping, and hunting as occupations. In addition to self-provisioning, much of this activity was market driven:

> Tons of catfish were shipped to the Midwest to be sold as tenderloin of trout. Large turtle pens enclosed herds of diamond-back terrapin being raised by the thousands for the restaurant trade.... In winter, market hunters regularly shipped more than 1,000 brace of ducks a week to New Orleans's markets. Oysters and shrimp were harvested by the boatload... In the 1920s,... Louisiana's marshes became North America's preeminent fur producing region. (Davis 2010:137)

Toward the end of the nineteenth century, prices for fur increased, resulting in a boom in the fur industry and an expansion of trapping. This was particularly significant in Terrebonne Parish. Ruth Underhill, a former official with the US Bureau of Indian Affairs, wrote a report on trapping among the Native American population that described the situation around the turn of the twentieth century:

> Trapping is the most remunerative [of occupations], but takes place only for three months when hunting is permitted, November, January, and February. At this time a whole family moves into the marshes, having their shanty boat towed, if they have one, or otherwise camping. The man has about two hundred traps which he visits, the women and children clean the skins. Until 1924, anyone could trap anywhere, and the bayou people made a good living from their winter's work. Just after the war, when fur prices were high, some cleared three or four thousand dollars. The swamp land, however, was private property (bought from the levee district of Atchafalaya in 1895) and the owners, noting the high fur prices, decided to charge for trapping permits. This reduced the profits of all who had not saved enough to buy permits which, of course, included the Indians. Permits were bought by middlemen who either hired trappers at $2 a day or took a percentage of the furs, leaving the trapper to sell the rest. Most of the Indians work on a percentage basis and are in decided need of help with their sales. (Underhill 1938)

The explosion in fur prices led to some of the first precise mappings of the region as businessmen began to try to lease specific areas for trapping rights. Often local trappers found themselves having to pay fees to hunt or trap on property they had habitually used (Davis 2010). Conflicts over access to land for trapping of muskrats led to intense struggles, known in nearby St. Bernard Parish as "the trappers war" (Gowland 2003). The trapping boom, however, would not last long.

Throughout the nineteenth century, the economies of Terrebonne and Lafourche were divided. Residents of the northern, farmland portions of the parishes were engaged mostly in sugarcane, cattle, and other agri-business production. Residents in the southern, wetland areas were able to mostly self-provision, supplying the majority of their own household food needs from what they grew, caught, hunted, or exchanged with neighbors. These down-the-bayou residents also participated in the cash economy. People engaged in logging and trapping, as well as fishing, hunting game, or growing vegetables, all to sell in markets in Houma, Thibodaux, and New Orleans.

The Twentieth Century

The life patterns established in the eighteenth century continued seamlessly into the early part of the twentieth century. For at least the first few generations, people living in the lower part of the parishes continued mostly as trappers, fishers, and small farmers. Trapping remained particularly lucrative. A local newspaper, the *Houma Courier*, estimated in a story published in 1925 that Terrebonne Parish produced more fur than any other parish or county in the US, and helped Louisiana lead the rest of the US in fur production. In another story published in December of 1925, the newspaper said that $350,000 had "passed through three Houma banks last week for the purchase of furs, breaking the record for any one week" and went on to estimate that more than $2,000,000 worth of furs would be caught in the parish that winter (Ellzey 2006). Esmiel DeHart, who was born in 1931 and raised in lower Terrebonne Parish, started trapping when he was a child and described the process:

> Since I was 10 year old, I'd go trapping muskrats.... We would go trapping during three months in the winter. We'd go to school until the first of November and then get out of school from November until the end of February. And we had to bring our homework and everything on it to do out at the trapping camp. Because we would only come in every month to get groceries. We would mostly live on the land, eating ducks and rabbits and trapping muskrats and everything—minks, otters, you name it. Now, I was raised up like that until I was age 16. (McGuire 2008)

Most furs were sold to dealers. But as Mr. DeHart's narrative shows, working in the fur trade went along with living in camps and self-provisioning. Often the whole family would participate and eat whatever they could hunt or catch for the winter. In this annual cycle, people would have a more permanent house slightly up the bayou and a seasonal camp farther down the bayou or in the actual marsh. By the Great Depression, most of the mink, otter, and muskrat in south Louisiana were depleted, and prices fell as the world economy crashed. People continued to trap, but the occupation was not as lucrative or popular. However, the seasonal pattern was well established in the culture. Native Americans—including elders and relatives of the contemporary Point-au-Chien Indian Tribe, the Isle de Jean Charles Band, the United Houma Nation, and the Grand Caillou-Dulac Band—continued to trap into the 1980s (Dardar 2007).

This decline of trapping would have created a significant economic problem for wetland households, but as trapping was becoming less lucrative, the oil industry arrived in the area and began hiring local workers. By 1901, Terrebonne Parish was dotted with natural gas wells (Cenac 2017). Even greater changes were under way in the 1920s when Texaco began drilling wells in Terrebonne and Lafourche Parishes. At first, Texaco relied mostly only on their own workers ("Texiens"), but soon job opportunities opened up for some local people. In the 1940s, both the entry of the oil industry into southern Louisiana and World War II brought tremendous changes to the area as roads were built and returning soldiers were offered new educational opportunities (McGuire 2008; Sell and McGuire 2008). For example, Pat Landry was the sixth generation in his family to live on Grand Isle. His father and grandfather were oystermen. When the time came for him to decide whether he would finish high school and go to college or whether he would take over the business, he chose school:

> He wanted me to take over. This was 1954. This was hard, back-breaking work. But he loved it. He told me, "Pat, I tell you what, if you want the business, take it over from me, I'll stay in the business and we'll build another big boat." And we were really going to get a nice boat. And it will be yours. But I had worked there as a little kid, and I said, "Man, there had got to be a better way to do that." It was cold and muddy and the gnats would get to you and, man, this is not for me. (Sell and McGuire 2008)

The oil industry, along with expansion of the shrimping industry, in the 1920s and '30s were driving forces behind the additional push of Anglos and Acadians into the lower bayous, crowding Native American space and resources. The United Houma Nation describes it this way:

> With the encroachment of French settlers, the Houmas began migrating south until they reached the lower reaches of coastal Louisiana. Because the land

was located along the flood plains of the Mississippi River, it was considered uninhabitable by most settlers. The Houmas were able to live peacefully off of the land, which provided all of their nourishment. Tribal members were traditionally farmers, fishermen and trappers. With the discovery of oil and gas in the 1930s, Houmas became vulnerable once again. (United Houma Nation n.d.)

While oil brought even more people down the bayou, there has been no time since the 1800s when the bayou land—even the lower bayou reaches—have not been used for cash crops and extensively exploited as a resource. Plantation culture continued well past World War II, with plantations growing not only sugar but oranges, pecans, and even Easter lilies (Cenac 2017). Many of the communities down the bayou, such as Klondyke, are named for the former plantations that sat on the lands. Rather than empty land, the shores along the bayous were dotted with facilities for lumber processing and shipping, sugar mills, syrup-making factories, oyster-packing factories, and shrimp-drying platforms (Cenac 2017). The land down the bayou today is riddled with hundreds of miles of canals, with early canals built by lumber companies and business owners seeking to ship seafood and vegetables more quickly to the New Orleans market (Davis 2016). Later, canals grew exponentially with oil and gas exploration.

While wetlands residents (including Native Americans, Acadians, Anglos, and African Americans) continued to enter more full-time wage occupations through the twentieth century, they also continued to practice subsistence activities. The characteristics of new types of work, like the week-on-week-off rhythm of offshore oil field work, provided a way for many people to remain involved in regular subsistence activities, like shrimping (Marks 2012), particularly in oil boom times when workers were always in demand. In his oral history, Magnus Voisin elegantly explains how the oil industry and subsistence coexisted: "In the season, I shrimp. When the season closes, I'd go to work in the oilfield" (McGuire 2008).

4

HERITAGE, IDENTITY, AND PLACE

How to Make Jambalaya

Lora Ann Chaisson has worn many hats over the years, including working with the Inter-Tribal Council of Louisiana, the US Department of Labor, and national tribal groups. She served as a member of the United Houma Nation Tribal Council for fourteen years, and in 2022, was elected Principal Chief. She's also a basketweaver, beadworker, and makes both alligator and alligator gar jewelry. Before the COVID-19 pandemic, she traveled to festivals around the US, and served as a representative of the Houma internationally. Almost every week, however, she drove to Isle de Jean Charles to give her dad, Theo Chaisson, a day off from managing his marina. "Other than that, I take care of the family." And she has lots of family along several bayous. "Dad's mom is from the Island [Isle de Jean Charles]. I have family in the Dulac area. On my mother's side my grandma was from Little Caillou—a large Indian settlement there. And her father [my great-grandfather] was from Dularge. Mother's father was from the Dulac/Grand Caillou area. My grandfather spoke the language—Houma, not French."

Lora Ann also has relatives in Houma. "Two solid streets were Indian—that was family." Her family all hunts and fishes, even the ones who work offshore. "They kept their shrimping licenses!" Her grandfather taught them all to duck hunt with hand-carved and hand-painted decoys. In 2013, Lora Ann was living in Pointe-aux-Chênes, in a house raised eleven feet off the ground, near a cluster of cousins, right beside where she liked to fish, and she invited Helen to spend a few days at her house. The following is from Helen's field notes.

We are sitting at her kitchen table, and she's showing me how to make jambalaya. Lora Ann is an accomplished cook who regularly cooks for family, friends, and for community events. She has often shared her culinary heritage at the New Orleans Jazz & Heritage Festival, where she has demonstrated how to prepare dishes such as white beans, maque choux corn, and filé gumbo. We are sitting in her bright kitchen, talking and chopping. Lora Ann is sautéing the onions. There's little I enjoy more than acting as prep cook to someone who knows what they are doing around the kitchen. So, I offer to chop some of the seasoning, onion, bell peppers, but Lora Ann does most of the work. "I don't write down any of my recipes." So I know this is a rare chance to learn from a master.

"You don't want no white jambalaya," she says, as she shows me how dark the onions are before she adds the bell peppers. "I use cream of celery, French onion, 1 large yellow onion, a can of Rotel, ¾ stick of butter, bell pepper, 2 cups of rice (washed), whatever meats you want to put in there. You throw that shit in the oven. That's fast. That's easy. It's good."[1]

Her language is salty, her manner is no-nonsense, and her cooking is delicious. Lora Ann used herbs too, but she didn't measure them, and I didn't write them down. She puts aluminum foil on the baking pan and slides it in the oven. I ask what temperature she has the oven on. "370° or something like that." The trick is to use the oven, to make the rice cook evenly, without burning the bottom, just like Lora Ann does.

✦ ✦ ✦

The last chapter considered the historical forces that shaped the economy of the bayou region, explaining that many populations of people settled in the Terrebonne-Lafourche area since European contact, including groups ranging from Filipinos (who developed shrimp drying) to Italians (who worked as sugarcane laborers) and Africans and those of African descent, who initially arrived as enslaved labor. The areas down the smaller bayous, where we did much of our fieldwork, have been dominated by two major cultural groups—Native Americans and Cajuns. Despite encroachment on land and growing participation in the cash economy, subsistence practices remained firmly in place for most Native Americans and Cajuns in the lower bayou region throughout the twentieth century. The snapshot of Lora Ann Chaisson that opens this chapter introduces this community scholar a little more closely, to show how her experience of identity is grounded in place. First, the roads she drives on are lined with people she knows and, much of the time, people she's related to. She once joked, "I would have to do a background check before I date somebody!" Second, her relatives and friends, like her, earn at least some of their income from the land or water. Her dad operates the marina; she makes alligator jewelry and palmetto baskets. Their community has turned the local foods into their mainstays. In her jambalaya recipe, she signals the local by saying, "whatever meats you want to put in there." For her that could mean anything from shrimp to redfish to duck or venison sausage, alligator, oyster, crabs. She makes the filé for her gumbo from sassafras leaves, just like her parents, grandparents, great-grandparents, and as far back as the stories go. "You can't buy that [pure sassafras] in the store." For Lora Ann, being Houma and cooking in a particular way with ingredients harvested by people she knows connects her to her Indigenous relatives. Her stories of being Native American are tied to her narratives of place. In this chapter, we take a closer look at identities (or identity labels) that have become linked to a distinct sense of place (specifically the bayous and waterways) and to deep hunting, fishing, gathering, and gardening traditions.

In coastal Louisiana, as in many parts of the US, the way people think about their identities doesn't always align with census categories. Labels such as "White, Not Hispanic" do not capture the complexity of identity for people down the bayou. We will look more closely at three groups—Native Americans, Cajuns, and African Americans. Then we will talk about how these identities, while distinct, can sometimes overlap, as people have complex ways of seeing their heritage. Our focus on these specific identities comes from the groups that emerged as dominant in our fieldwork. However, we want to be clear that we also know that people with many other identities practice subsistence in the bayou region. We were particularly aware that we documented very few people from the Latino, Vietnamese, Laotian, or Filipino communities.

Native American

Native Americans in the region, from the 1830s until the 1970s and 1980s, often submerged their Indigenous identities choosing other labels because of the stigma associated with being Indian. During her early 1990s fieldwork, white people themselves living "down the bayou"[2] explained to Shana that there were "no Indians" living in Terrebonne Parish. She remembers one neighbor saying, "Don't let them fool you. They aren't Indians." Years of hard work by Native American groups to reclaim their identities has increased awareness and brought some official recognition. There are now several state-recognized Native American tribal groups in Terrebonne and Lafourche Parishes, including the members of the Biloxi-Chitimacha Confederation of Muskogee (Bayou Lafourche Band, Grand Caillou Dulac Band, and Isle de Jean Charles Band), the Pointe-au-Chien Indian Tribe, and the United Houma Nation. In 2018, several of these groups were in the process of petitioning for federal recognition or were appealing a recognition decision.[3]

For many Native Americans in the region, their history of subsistence practice remains key to their identity. In his history of the United Houma Nation,

Daniel d'Oney (2020:57) says that it is "impossible to understand the Houma without understanding their relationship to water," not only as a primary food source, but as a way in which their culture has a deep sense of place. We think of Lora Ann's discussion of where her relatives live: two grandparents from Isle de Jean Charles, one from Bayou Little Caillou, one from Bayou Dularge. All waterways. Echoing Monique Verdin's idea of the swamp as the "safe refuge," d'Oney writes that the Houma saw their relocation as "hope" (53) that they would not once again be pushed aside. Their hope was that they now lived in an area so undesirable that they would not have to worry constantly about being pushed out of their homes. "Between 1762 and 1850 the Houma created a new world—physically and culturally—for themselves along the bayous of southeastern Louisiana" (143). Creating this new world meant transforming their farming and trading economy into what d'Oney calls "a hunter-gatherer lifestyle" (95), with seafood a dominant food source. They developed netting methods, learned to use local plants for food and healing and how to harvest oysters and clams, passed down how to grow food, and navigated waterways to share and trade. Archeologist T. R. Kidder (2000:20) says the Houma "remade land in fact and in the mind." In this way, the Houma economy and culture, both based on bayou resources, became entwined. Scholar Tammy Greer, herself a member of the United Houma Nation, says that land, plants, and culture are linked:

> These plants are part of our culture. They're the cane and palmetto and white oak in our baskets, the supplejack in our basket handles, the cane and elderberry in our blowguns, the arrowhead of our arrows, the history in our stickball sticks and rabbit sticks, the corn in our hominy, the sassafras leaves in our gumbo. Our medicine, our food, our building materials, tools—the plants are every aspect of our culture, and we can't forget that. (Botanica 2020)[4]

And of course, the plant life is a product of the waterways environment. Houma historian T. Mayheart Dardar explains the deep connection this way:

> Houma existence is tied to the bayous that are the foundation of our community. Our very identity is tied to that sense of place recognizing that we are a Houma from Bayou Lafourche, or Bayou Grand Caillou, or whatever bayou the family is from. Most Houma families are directly involved with the commercial fishing industry or are no more than a single generation removed from it and only the crash of the fur industry in the 1980s separated most families from that form of livelihood. (Dardar 2007)

Among Indigenous communities seeking recognition for their right to exist or for those continuing to fight to be seen and to be recognized, subsistence histories and heritage tie them to place and to their hold on the future as a people.

Cajun/Acadian

The story of the Cajuns recounts how the British forcibly removed the French-speaking Acadian farmers from their homeland in Nova Scotia starting in 1755—an event known as *Le Grand Dérangement*. More than ten thousand Acadians were dispersed as refugees to the Caribbean, the British Isles, the eastern United States, and France. Historians estimate that as many as five thousand Acadians died at sea or of starvation. Starting in 1785, Spain paid for ships to resettle Acadian exiles in their new colony, Louisiana, and over the next ten years more than four thousand made the trip (Brasseaux 1987, 1991, 1992).

The story is an amazing tale of a people of great tenacity. And, yet, back in the early 1990s, only a few of the people Shana interviewed down the bayou (including French speakers) who identified as white also saw themselves as "Cajun." Older people seldom used the word or corrected her if she suggested the label. One neighbor, Elsie Lejeune—Lejeune is one of the Acadian surnames—said, "They tell me I'm Cajun. But I'm French." Others embraced it. The late Claude Bourg, who ran a small store in Montegut where everybody stopped for morning coffee, said, "I used to say I was a Frenchy. But people came and taught us about the Cajuns. It's something to be proud of." Both of them were noting that in their childhoods the word "Cajun" had been used as a slur, a way you indicated that someone was poor and uneducated. For years, they had identified with the language, but not the ethnic category.

As Claude Bourg noted, that reluctance was changing. In the 1960s and '70s, in areas near Lafayette, Louisiana, there were movements afoot to reclaim the French language and Cajun identity:

> While Cajun pride soared in southern Louisiana, mainstream America "discovered" this unique culture in its own backyard. Reagan-era yuppy-ism, with its emphasis on conspicuous consumption of the new and exotic, fueled a veritable Cajun fad in the 1980s. (Bernard 2011)

This cultural revitalization effort spread from the Lafayette area to the bayou region. By the late 1990s and early 2000s, the label "Cajun" was common, if not ubiquitous. In 2018 a billboard on Highway 90 leading from New Orleans to

Houma proclaimed, "Now Entering the Cajun Bayou." In fact, some scholars have noted that the "Cajunization" (Trepanier 1991) of south Louisiana has lumped together many heritage French speakers and other migrants.

For Cajun identity, some people look to immigration rolls from the 1700s to define their heritage, searching to see whether or not their surname was on one of the boats that carried deportees. However, when many others label themselves "Cajun," they are pointing to lifestyle, where they live, values (like spending time with family and friends), activities like hunting, fishing, or crab boils, or even just French last names (many that were not originally French—like Labranche or Plaisance—or many that had had other French origins—like Champagne). As one person in the documentary *Finding Cajun* put it: "I have the last name. I hunt. I fish. That's enough to be a good Cajun" (Rabalais 2020). In another example, Mike Tidwell in his book *Bayou Farewell*, hitches a ride on a shrimp boat headed down Bayou Little Caillou, from the small town of Chauvin to Cocodrie (the end of the road) with Charlie Broussard, who spends much of the trip defining what makes a "Cajun." He says being a Cajun is having "a knack" for living off the land, navigating the bayous and marshes, being outdoors, and loving family, friends, and good food. He explains: "For me, it's when that old morning sun comes rising over my boat deck and the boat's covered with a ton of shrimp, and then I have me a big bowl of jambalaya with the guys at the shed, and we all everybody got money in the pocket. That, to me, that's bein' Cajun!" (adapted from Tidwell 2004:57).[5]

The key elements Broussard points to—being on the water, harvesting their own food, prized dishes, and sharing with friends—come up again and again as key markers of Cajun culture (see Rabalais 2020).

African American

Currently, the lower portions of the two parishes have a relatively small population of families who identify as Black or African American; however, that was not always true. According to census records, there were sizable African American populations down the bayou until the 1930s and 1940s. The years from the turn of the twentieth century through World War II saw a major exodus of African Americans. The reasons are mixed, and include the search for greater economic opportunities, mechanization of agriculture, and a growing oil and gas industry that operated with a bias against hiring racial minorities in Texas and Louisiana—limiting the types of jobs that could be held by African American and Mexican American workers while giving the best jobs to white men (Dochuk 2019; Pierre-Louis 2021; Priest and Botson

2012). Also, importantly, both Lafourche and Terrebonne Parishes were sites of intense racial violence in the period after Reconstruction and during the early Jim Crow era. To understand this, we examine the Thibodaux Massacre, one of the most violent racial mass killings in US history.

After emancipation, plantation owners worked to keep formerly enslaved African American laborers subjugated. Many continued as sharecroppers, often living in former slave cabins, and were paid in "scrip" that could only be spent at the plantation store, perpetuating a new Jim Crow form of bondage. Vagrancy laws were used to prevent any unemployed African Americans from remaining in the area, a practice that proliferated throughout the South (Bell 2021; DeSantis 2016; Schermerhorn 2017; Scott 2005). Despite this, working with organizers from the Knights of Labor, cane laborers organized a major work stoppage to protest conditions and payment practices. With more than ten thousand workers across four parishes participating, the Sugar Strike of 1887 was one of the largest labor actions in southern history, and one of the few that was interracial, as immigrant Italian cane workers joined with the majority African American strikers (DeSantis 2016; Scarpaci 1975). The strike began on November 1, 1887, during a critical time for rolling and pressing cut sugarcane to extract the sugar and lasted three weeks (Schemerhorn 2017). Though some of the smaller planters negotiated with the workers, the large plantation owners refused and fired the union leaders, who then had to leave their lodgings, and hired strikebreakers. All-white militias were mobilized to face the striking workers, led by former Confederate general P. G. T. Beauregard, and in Thibodaux, martial law was declared. On November 23, the militia began shooting in Thibodaux. "Bodies were dumped in unmarked graves while the white press cheered a victory against a fledgling black union" (Schemerhorn 2017). The massacre that resulted was one of the largest mass killings of African Americans in US history (Equal Justice Initiative 2017). The actual number of people killed varies by source—estimates range from twenty to fifty—but we know that despite the interracial nature of the strike, all fatalities were African American (DeSantis 2016).

This event was an outlier, but only in scale. Researchers have documented dozens of other lynchings in the bayou region between 1880 and 1940 (Equal Justice Initiative 2017). This campaign of terror had long-term implications for the demographics and subsistence practices in the region as African American workers and others seeking greater freedom, safety, and economic opportunities left the area for New Orleans and beyond.

The city of Houma has one of the largest concentrations of African Americans, with the American Community Survey estimating that in 2020, more than 23 percent of residents identified as Black or African American.

But, to this day, African American residents who live down the bayou are often congregated in smaller communities rather than scattered throughout the lower part of the parish. For example, in Terrebonne Parish, a community known as Smithridge, located between Houma and Chauvin, remains predominantly African American. That community dates to the aftermath of the Civil War and is linked to land that was once part of a sugarcane plantation but was settled by two brothers who purchased the land and farmed it (Cox and McQueeny 2022). The few dozen families remaining in that community comprise the majority of residents who identify as Black or African American in south Terrebonne. Bobtown, a community farther down the bayou than Smithridge, has a similar story, with the formerly enslaved Bob Celestin buying up a parcel of a former plantation in search of a location that could be a haven for a Black community (Charles 1997). Although those communities have dwindled, their legacy lives on. Houma resident and community activist Cherry Wilmore points to these communities as touchpoints for the African American people in the region: "You say Chauvin and Dulac; I say Smithridge and Bobtown" (Wilmore 2022).

In spite of the region's difficult and violent histories, many African Americans, whose ancestors migrated out of Terrebonne and Lafourche, return to the region for family reunions, to tend to family graves, or to spend a day fishing. During our roadside fishing surveys, we met a number of African American fishers who drove one or two hours to spend the day by the water and to catch their supper, enjoying the tranquility and beauty of the place. In addition, the African American history of Terrebonne and Lafourche Parishes is well remembered in some sections of New Orleans, which were populated by migrants from the sugar-producing parishes of the lower Mississippi Delta. Many brought with them aspects of their rural heritage: the habit of growing food crops, keeping a garden, hunting and fishing in nearby lakes and bayous (such as Bayou Bienvenu in the Lower Ninth Ward), and sharing with neighbors. In an oral history recorded by Mr. Philip Ganier, we get a sense of how relocated African American communities transmitted aspects of the bayou subsistence practices into the city: Mr. Ganier said, "Everybody knew everybody. And what I like about it is . . . we learned what lagniappe was."[6]

A similar pattern was found among DC area subsistence fishers by Shirley Fiske and Don Callaway (2020:175):

> We were surprised by the strength of what we described in our report as a "southern rural subsistence tradition," especially among African American and Native American/Piscataway fishermen and fisherwomen who fished the Potomac and Anacostia. It was a coherent expression and dedication to fishing, traditional food ways, and self-provisioning over generations. [. . .] Many of the

fishermen we interviewed had been sent by their parents during the summer school break to grandparents or aunts or uncles in the South—South Carolina, North Carolina, Georgia. During the summer they learned to fish, to hunt, and to cook the foods they caught. They retained their foodways and are passing them on to their children and grandchildren now in DC. As one man put it, "Fishing is in our family legacy; our heritage."[7]

As Fiske and Callaway (2020:175) reported: "Some fishermen were very articulate about coming from a subsistence legacy in the South that was once a way of life. We inquired about whether one fisherman's father's parents fished (in South Carolina), as part of the oral history, and it led to the following description of living in the South:

> I suspect that most of the people from the South and most of the people who were living on the edge of poverty knew how to fish, because fish, fishing, was a source of food. *Nobody fished for fun. I hadn't heard of catch and release until I got to be a grown man.* Nobody caught and released at that particular time that I know of. Now, there was a lot of sharing. People would catch fish and give some to the man next door. In fact, my friend who comes down here to fish, he fishes, he catches 130 perch and he gives some to the barber and some to the lady next door and so forth and so on and so on. At that time, people fished for food. There was no fishing for fun or catching and releasing. This was serious business most of the time. It was a way of life. (Fiske and Callaway 2020:175, emphasis added)

Layered Identities

These racial, ethnic, and identity labels we have discussed are not exclusive, nor do they have firm boundaries defined strictly by descent. For example, a person who is a member of a Native American tribe may also explain to you that he's Cajun. In one case, a man who identifies as Native American told us that when he was younger, he thought of his family as Cajun, but as he learned more about his heritage and community, he came to know and value his kinship with his Indigenous relatives.

"Cajun" is just as complicated. The label is an Anglicization of "Acadian" or "Cadien" and became more common after the Civil War, allowing poor French-speaking whites to evade identification as nonwhite (see Rabalais 2020). Despite "Cajun" being commonly associated with whiteness, some African Americans also claim the label as their own.[8] And not all who identify as white consider themselves Cajun, sometimes because their ancestors weren't French-speaking or haven't been in the region for generations, or because they don't identify with the associated lifestyle.[9] Sometimes they select another label. One of our community scholars, Wendy Billiot, married a Native American man and lived down the bayou for more than forty years, making her living, in part, as a fishing guide. However, because her childhood was spent in north Louisiana, she always notes that she's a "bayou" person, but neither Cajun nor Native American. Also, just to complicate identity markers more, although French names and the French language remain key markers of Cajun ancestry, the Native Americans of the bayou region are the people actually more likely to still speak French as a home language (Rabalais 2020), and many surnames are shared by people who identify as Cajun and Native American.

There remains a stubborn Black-white divide. Most of the African Americans in Smithridge, for example, do not consider themselves Cajun, and only a few consider themselves as also Native American. However, some of their community members see their culture as embodying similar values and lifeways—good food, connection to family, living off the land. A community leader from Smithridge, Effie Bennett, explained that her community also values good food, neighbors, and sharing, and has found creative ways to live off the land for generations.[10]

There is also a growing recognition that the communities down the bayou have long intermarried. Filmmaker, photographer, and author Monique Verdin, whose family are Houma, tells the story of how her father would frequent African American bars in New Orleans and take up a Creole identity to blend in (Verdin 2012). The term *Creole*, at one time, was frequently used to describe biracial (Black-white) people, but is growing as a term to describe multiracial people, including people who also have Native American heritage (Jolivette 2007). Noting that Native people were the original residents of this region and that intermarriage and sexual relations were common between Native people and the incoming groups, Jolivette writes that Louisiana has "a long history of Indian-Creole, Indian-French, Indian-Spanish, and Indian-African relations" (2007:6).

In sum, when you meet a person from down the bayou, you cannot always assume a particular identity. None of these layers means the labels are empty (or meaningless) nor are they necessarily interchangeable, although they might be shared. Throughout the rest of the book, we use the labels people use to describe themselves (how they self-identify), with the understanding that these identities have been forming and reforming for generations.

In other contexts, we often heard people use a general word "heritage," as a shorthand for asserting a connection to place and to local traditions, avoiding any specific cultural, ethnic, or racial identity and, instead, focusing on shared activities and values. Decisions around hunting and self-provisioning were often couched as a form of heritage identity. Even if the number of ducks shot per year, for example, makes up a small portion of a family's food budget, hunters often seem to identify duck hunting with self-provisioning and subsistence heritage. One student, Trey, wrote in his essay on duck hunting that the hunters "see it as something to teach and pass on to their children and grandchildren as it was learned from their ancestors." Contemporary duck hunting, thus, is subsistence heritage.

One history of the region claims, "Few residents today are without some sort of boat" for fishing (Ditto 1980:37). While that claim might be exaggerated, fishing, crabbing, shrimping, hunting, frogging, gardening—and more—have long histories as both subsistence and commercial practices in Terrebonne and Lafourche Parishes. These long histories ensure that such practices are enmeshed in local constructions of place and identity. Most people whose families have lived down the bayou for any length of time have their history grounded in subsistence practices, which makes sustaining these practices precious to people. And, when group labels are so intimately connected to lifeways, when those lifeways are under threat—as they are now from environmental changes and land loss—the people face an additional existential threat to their identity.

5
FAMILY, COMMUNITY, AND FEASTS

Crabs = Happiness

Jacob was a freshman when he wrote "Crabs Under a Microscope," for his honors composition class in 2011. In this snapshot, he argues that a close look at crabs and crab boils is one way to map his family relationships.

    As I pull into my grandma's driveway, I can already hear the commotion and loud screaming coming from my family inside. As soon as I step through the doorway, a rush of sights, smells, and sounds enter my body. My whole family is sitting at a long plastic table, and I already know crab season is here.[1] Grandma kisses me on the cheek and tells me to get a tray and sit down. My grandma is viciously beating her crab with a knife to try and crack the shell. Finally, when she succeeds, she opens the crab and picks the meat to give to the children who cannot peel for themselves. Down on the other end of the table, my older cousins compete to see who can peel their crabs the fastest. The whole table is exploding with noise from every direction. A few of the children are singing and dancing in a circle adjacent to the table. The smell of freshly seasoned and spicy crabs fills the air. The smells are so strong that they burn my nose, and my eyes begin to water. After everything is cleaned up, my whole family piles into several cars to go get ice cream from a small shop in Napoleonville. My mom says that it is nice to have something sweet after eating the salty crabs. Our night then comes to an end, and we go to bed with full stomachs.
    Family crab boils are a very familiar scene to me. I have participated in them my whole life. In Louisiana, crabs do not only serve as another source of food. They serve as the backbone to many relationships and family ties.

I interviewed David, my best friend's father and a barber in Thibodaux. He also receives and distributes crabs to several people in the Thibodaux area. An old man by the name of Sterling catches the crabs in Flat Lake in Morgan City. Sterling is a 70-year-old man who learned how to live off the land from his father and became a hunter, trapper, and crabber. He always had another job to support his family in case times became hard with the crabs. He is a very popular man in Morgan City, and he sells his crabs to show off his hard work. Once Sterling catches his crabs for the week, half are sold to distributing companies and sent throughout the United States, and the other half he sells to close friends. Every Monday, David travels up to Morgan City to get the crabs from Sterling. He said that the Monday trip is always an adventure. He sometimes brings a friend to ride with him. He said the conversations usually consist of crazy childhood memories and stories. David told me that he thinks the ride is sometimes more fun and important than actually getting the crabs themselves. Once David obtains the crabs, he throws a huge crab boil for his family. He only sells his crabs to his close friends when he has a surplus.

Luckily, my family and I are very good friends with David, and the crabs get passed further down the chain to us. Crabs have played an important role in relationships between people and families. They are sacred to the people of south Louisiana. Without the crabs, David and Sterling might think very differently about society. It would also affect the way they act towards their family and friends. Most importantly, if not for the crabs this amazing relationship would have never been created between these two men. They might have never even known each other.

Besides affecting relationships and conduct in people that catch and distribute them, crabs also are very important to family life. Just the simple task of boiling crabs brings the whole family together. Once the crabs are caught and brought in, the whole family knows it is time for some quality bonding and great food. It is a time to put behind all of the chaos and distractions of life and just focus on what is most important, and what a better way to do it than to celebrate with the catches made off the land. This ties together Louisiana's culture of living off the land and close family bonds. The crabs are taken from the land, they affect how many relationships are formed, and they influence family life.

✦ ✦ ✦

In his essay, Jacob draws a straight line from the crabber in Morgan City[2] to his family's connectedness. As with the essay by Rory and the portraits of Jerome and Glynn, this snapshot of a family crab boil points to the key role of feasting as social glue. Louisiana has been labeled as a "sticky state" (Pew 2009), meaning that residents are more likely to be born here than to have migrated here. In fact, in 2020 nearly 75 percent of Louisiana residents over twenty-five were born in the state, and the four counties in the US with the highest "native-born" population are all in south Louisiana (Maciag 2019). No doubt there are many bonds holding people together, such as strong religious ties, but during our research it became apparent that for many people subsistence practices play a key role in these tight family ties, particularly given the deep history of wild harvest in many people's families. In fact, some people defined hunting or harvesting in just those terms. At the Chauvin Folk Festival in 2013 we talked to one woman in her thirties, who grew up hunting with her dad, and now often goes fishing with him, and she said, "Look, hunting is all about family. That's it."[3]

We might be tempted to leave it there, but instead, we draw on the power of ethnography to show *how* these connections are made through specific examples. In this chapter and in chapter 6, we explore several streams that make up subsistence, including how such practices are linked to community and family and contribute both to individual and group well-being. We also look at how hunting and fishing are key to what anthropologists call *enculturation*, or the process of shaping a human person to a specific culture. In this case, we specifically mean raising a child to understand what a community values, what attitudes are ideal, and what behaviors are respected. In the conclusion, we connect these ideas of close family links and proper enculturation to larger theories about the anthropology of happiness.

Harvesting to Facilitate Family Gatherings

Families in south Louisiana are famous for their large size and for their closeness. The documentary *Happy*, a 2011 film made by New Yorker Roko Belic, presents the Blanchard family of Louisiana as exemplary of people who have found the elusive state of happiness that so many people seek. In the documentary, Roy Blanchard Sr., a crawfish farmer who lives in the Atchafalaya basin, is shown first talking about the peace and serenity of the swamp, and then in a pirogue crabbing for a family crab boil. Next, the family gathering is shown, with Roy's siblings and their spouses feasting on crabs and laughing and telling jokes. Roy echoes what Jacob says by saying that to have this weekly gathering without being able to catch the crabs for free, "you'd have to be rich." The film clearly wants you to see that these close family ties, this routine togetherness, combined with a love of nature, are what enable the Blanchards to be happier than most Americans. And people got the message. On her blog, caughtncooked.com, living-off-the-land advocate Paige Yim writes about the scene from the movie:

> Crabs = happiness. It's true.... The Blanchard family sums up the key to happiness best—Live off the land, catch crabs and have tons of crab feasts with family and friends. Beer helps too. (Yim 2019)

In our fieldwork we found many others who made the same connection. Specifically, they connect their ability to harvest delicious, low-cost or free food with their ability to maintain family ties and close relationships. Those ties, in turn, allow them to have a sense of well-being.

Looking more closely at Jacob's essay about crabs, at one point he calls them "sacred" for the role they play in family life. "Just the simple task of boiling crabs brings the whole family together." Jacob's essay dwells on the sounds, smells, and tastes the family is sharing. He writes, "As soon as I step through the doorway, a rush of sights, smells, and sounds enter my body." And he writes about his grandmother picking out the crabmeat for small children while the "smell of freshly seasoned and spicy crabs fills the air." The saltiness of the crabs, he writes, makes their ice cream taste more delicious. The family are not only together, but also they are bonded by the sight, smells, and tastes of delicious food. In a part of the essay not included in the opening vignette, Jacob interviews his father, who says that crabs play a central role specifically because of their low cost:

> According to my father, crabs are more accessible to people in Louisiana than any other seafood. He says they can be easily caught in ditches and small canals. You also need very few of them to feed yourself and others.

Jacob's family members are not themselves crabbers, but buy the crabs at cost from a friend (who also sometimes gives them some). Jacob said this lowers the cost sufficiently that his family can afford to buy enough to feed aunts, uncles, siblings, cousins, and in-laws at least once a week—and sometimes twice a week—for several months. "Without the crabs, we would not have as many family gatherings," Jacob said in a classroom discussion about his research. One remarkable point to Shana was the utter lack of complaint in student essays about what appear to be almost mandatory family gatherings. These are college students who willingly spend most weekends with their families. In his essay, Jacob says the crabs are a signal that "It is a time to put behind all of the chaos and distractions of life and just focus on what is most important"—in other words, family. Jacob was not unusual among students we interviewed. In fact, we found that his attitude was the norm rather than the exception.

Although both Jacob's family and the Blanchard family identified crabs as the key to these gatherings, other low-cost harvested foods, particularly fish and shrimp, serve the same purpose. For example, our fieldwork included many days of drop-in interviews along favorite fishing sites, frequently Island Road, which connects Pointe-aux-Chênes to Isle de Jean Charles. Shana's field notes from a hot August afternoon include a husband and wife who doubt they could feed their regular crowd without their fishing trips:

> Their van is parked on the side of the road, and a cooler is between their two lawn chairs with canopies. The couple are retired and now live in Bourg (Terrebonne Parish). He self-identifies as Cajun and pronounced the local newspaper (the *Houma Courier*) with a French pronunciation. They come down the bayou to fish Island Road two or three times a week. Often they eat the fish that night, but if they catch a lot, they will freeze it and have a fish fry. This couple married after their previous spouses passed away, and between them they have 8 children, 25 grandchildren, and 14 great-grandchildren. They said the fishing supplies their supper about twice a week, and they have enough for a fish fry at least once a month, if not more often. So far that day they had caught one redfish and three drum.[4]

These fishing trips are fun and essential. The couple explained that it takes several trips to catch enough fish for the monthly gathering. They made it clear that having a freezer full of fish that they can catch for only the price of bait and gasoline allows them to feed almost fifty people every month.

Let's pause here a moment to take note of the fact that these three families—Jacob's, the Blanchards, and the couple fishing on Island Road—are all talking about frequent family gatherings. Two of these families are hosting at least twenty to twenty-five people each week and the other family is gathering together twice that many each month. In chapter 1, we saw Rory's family and friends also gathering at their camp in Grand Isle each week to visit and feast. These massive gatherings are pervasive. In fact, they are so common and so ingrained in the culture that undergraduates at Nicholls will often explain that they can't do an assignment for class until they get back from the camp or their grandparents' house.

Another student, Victoria Verdun, wrote an ethnographic essay on deer hunting showing how the harvested food facilitated gatherings. "Family gathering always entails one thing here—food." She explicitly connects the dots. "Hunting turns into food, food turns into gatherings, gatherings turn into bonding."

We have featured a good deal of undergraduate writing here, in part, because some readers might expect, as we did at first, that those who most valued the old ways were, well, older people. But this wasn't necessarily the case. We found it compelling how many people in their twenties asserted that these feasts based on harvested or hunted food were key to their family ties. In some of their analyses, the students found that people who are asked to talk about hunting and harvesting also end up talking about family. In the same way that Mrs. Dupré—from the oyster spaghetti story in the preface—filled her food logs with relatives and neighbors, the students found that people constantly referenced their family members and community connections in their stories of hunting and harvesting.

To look more closely at how this works, we will analyze an oral history with Lafourche Parish crabber and alligator hunter Joe Autin, who was interviewed by Annemarie Galeucia. Joe was one of many people who talked to us about not only his commercial harvesting career but also his recreational harvesting of deer and waterfowl. We took the bulk of one interview session that focused specifically on hunting, about six thousand words long, and filtered it for words that pointed to hunting or harvesting.[5] Deer came up forty-one times, fish thirteen times, and alligators thirty times. He mentions other game a scattered number of times and indirectly references hunting or harvesting other times. In those six thousand words, he explicitly mentions or references hunting or his commercial work with alligators almost one hundred times, which you would expect. We also filtered for words that indicated family, such as references to his children ("my girl," "my boy," "son," "daughter," "kids"), his spouse, parents, and other family members ("wife," "mom," "grandmother," "parents," and "daddy"). What emerges is that in an interview narrowly focused on hunting he mentions family in some way or other more than sixty times. And those sixty references are scattered through the interview, not clumped together in, say, a background biography section. Few paragraphs are entirely without a mention. So, in other words, the history he gives of his life hunting and harvesting is also a story about his family. Certainly, *family* is the running subtheme. Below is a portion of the transcript from one of the interview sessions, and it begins with Joe and his wife, Claudia, telling the story of how they started their honeymoon. They drove away from the church and ran into a deer.

> **Claudia:** You see that one that's crooked [*pointing to a mounted deer head*]? We killed that one.
> **Joe:** We ran over it on LA 24 going to Houma, and we turned around, and I threw it in the trunk of the car, and we brought it back.
> **Claudia:** $700 worth of work on the car. And it was so funny because my daddy was in his tuxedo, and he came to clean it. That's the story.

The Autins were among the many people who were concerned that our project would frame them as something like characters from the popular television show *Swamp People*.⁶ When Annemarie approached them about an interview, Claudia's first reaction was that they did not relate to people who "live off the land." She pointed out that they purchase most of their food in the store, like to eat out, and that Joe has a salaried job with benefits. But as she was telling this, she was cooking lasagna using deer meat sausage, the freezer was full of self-provisioned food, and the walls were decorated with mounted animals from hunts. And their photo box was pictures of family hunting. Both of their children, a son and a daughter, received their first guns at age three. "My daddy bought them their first guns. They were 20-gauge crack barrels,"⁷ Joe said.

We find the language corpus analysis interesting because we believe that it reveals a common pattern: hunters often wind up talking about family and friends. Like the woman at the Chauvin festival, many hunters tell you explicitly that hunting for them is about family and friends and others tell you implicitly, just in the way they talk, as the subject just keeps coming up. For example, we can look at an interview with Lafourche Parish resident John Serigny. John is a devoted duck hunter, who told us that most of his important friendships have been formed around his hunting. In one of his interviews, he explained that the purpose of hunting and the camp itself had become centered around keeping friends and family together, and it no longer mattered if he actually killed any game on the hunting weekends:

> So when we go to the camp on Thanksgiving and the others, which is usually the traditional beginning of the first split⁸ of the season, my wife and my two daughters would come. My friend John, who's been duck hunting with me since '69, he would bring his wife and his two daughters, another friend of mine, James, who's been duck hunting with me since the same time, would bring his wife and his sons, and we would fish and we would duck hunt. I'd bring the girls duck hunting, sometimes I'd bring James' boy duck hunting. He'd take them. I brought James' youngest daughter duck hunting. I mean, we'd go out on the youth weekend, and the kids would hunt. And it doesn't matter whether you'd kill anything, whether you catch the game bird or not. That's not important. It's the experience of going out there, riding in a pirogue, sitting in a blind, you know. My children—my oldest daughter is 37 years old—and she still talks about the time at the camp. And she loves going to the camp. That's where she got married.

Extended Family and Friends

For some people, hunting or fishing doesn't start with immediate family. We heard stories about how college friends, in-laws, or older cousins and uncles were the people who helped them fall in love with fishing, hunting, or gardening. Although he knew some hunting basics, deer hunter Richard Borne (who is featured in the next chapter) told us that his mother-in-law and father-in-law, who were trappers on Bayou L'Ourse, taught him to really learn to love the outdoors.

Lafourche Parish farmer Arthur Bergeron's father was a country doctor while Al Guarisco, whom we met in chapter 1, had a father who ran a restaurant and bar. Neither a country doctor nor a restaurant owner had much time to spend teaching their children to hunt, so both learned from older friends. We have already heard from John Serigny about how connected hunting is to family. Interestingly, while his family has always been a hunting family, and his father is known in Louisiana as a folk craftsman for his prized carving of duck decoys, John himself did not learn a love of hunting from his own father, who originally owned the camp with his friends.

> When I used to go hunting with those men, I really didn't enjoy it. I enjoyed duck hunting. But the rest of the time around the camp I just sat around. I'd just sit around the camp and fish, stuff like that. Basically, it was a long weekend for me. Then I went to college, and one of my friends who I used to play cards with in the game room and shoot pool with in the student union was a fellow by the name of John from Lockport. So he and I started talking about duck hunting one day, and he says, you have a camp? I said, yes, I got a camp in Leeville. He said, man, I'd love to go hunting. And I said, fine, man. He said, well, why don't you call somebody with a boat and invite them to go hunting? We can go in his boat. So, I called him . . . I said, "would you like to go duck hunting?" And he says, yeah. I said, can we use your boat? [*laughs*] And he says, yeah. So, we took off to the camp.
>
> We spent two nights out there. We really roughed it. I don't think we brought anything to eat even. We caught a few fish and fried some fish and stuff like that. But what we did do, we would sit up at night and talk about hunting. Talk about guns, dogs we had, and times we went hunting, and things that happened to us when we went hunting. And that's when I really fell in love with hunting. And I don't remember how many ducks we killed. . . . And those are some of the best friends I have. I have really good golf friends. And I have really good duck hunting friends.

John's story is told with humor and pacing that shows how he came to realize the value of the camp through his friends. He had the camp. One friend had the desire to go duck hunting, another had the boat they'd need to get out on the water. But what stands out in his story is the closeness that comes from spending time at the camp with friends. The suffering[9] (not enough supplies)

Figure 9. In a photo displayed in his house as a treasured memory, Arthur Bergeron (*left*) stands beside his cousin, Skip Bergeron, after a successful fishing excursion. Like many children, Arthur learned from teenagers who showed him the ropes. Image courtesy of Arthur Bergeron.

and the long nights of conversation. His father introduced him to the camp, but his friends made him love it.

Coming of age in a coastal community, even if there were no immediate family members to adequately enculturate you, then there were plenty of cousins, uncles, and friends. As with John, those peer relationships are important and can remain important throughout life. More than one community has a weekly or monthly dinner, where friends (usually just the men) gather to talk and show off their hunting skills (by bringing game or fish) and cooking skills. Those peer bonds can be seen in several stories people told us about camps and clubs. Arthur Bergeron has a picture hanging on the wall of his house showing him as a child proudly holding a catch of fish, standing next to an older teenage friend who taught him many of his hunting skills (figure 9). Arthur said he and the man remain good friends to this day from those bonds formed when they would spend the entire weekend camping, hunting, and eating only what they could catch or shoot.

Harvesting and feasting provide reasons for friends to socialize. For example, people flock to Louisiana's Grand Isle in summer not only because of the sand beach but also because the blue crabs are running—meaning they are plentiful along the beaches and shoals. Researcher Mike Saunders wound up meeting several groups of friends when he picked one day in August 2012 to crab at Grand Isle. One group of five retirement-age men and one woman, from places ranging from Lafayette, Baton Rouge, and Breaux Bridge, had all met up at a central point and driven down together (about two to three hours) to Grand Isle. Here are Mike's field notes:

> They were old friends who gathered together whenever they could to fish, at least once a year, making a trip like this, to stay on Grand Isle at a camp of a friend. When I asked if the crab run was the reason they were here now, they said, "No, not the main reason, just good timing." They explained that they tried to get together in the general time period surrounding August. They were in particular there for the crabs (and other fishing) because they were all "big crab-eaters." And tonight, they were going to "eat the shit out of them."

Their irreverent language suggests that for these longtime friends, crabs may not be sacred, but they sure provide a great occasion to get together. In either case, crabs are on the table, fueling social gatherings.

Passing It On: How to Raise a Good Person

How do you raise a good person? For many people we interviewed, the natural unit for transmission of the hunting and harvesting knowledge is the family. Sometimes the unit is also friends or peers (age-mates), but most people emphasized the family as key.

Fieldwork with Richard Borne show him and his friends teaching a grandson how to clean a rabbit. When Mike Saunders went on a rabbit hunt with Richard, he noted that after the hunt, they returned to the trucks and all the adults sat around and watched the grandson skin the rabbits:

> Back at trucks, we sit while grandson cleans his rabbits. He knows the basics but keeps getting advice from those gathered around but takes it well without looking too embarrassed. They also tease him about the amount of grass sticking to the cleaned rabbits, saying that they don't need a salad with it, it already has one and also don't need to garnish it with parsley.

Richard taught all his children and grandchildren, both girls and boys. For instance, his granddaughter Haley was also on that rabbit hunt (figure 10).

Figure 10. Richard's granddaughter Haley as a teenager with a rabbit she shot. Photo courtesy of Richard Borne and Haley Metzger.

Youth who have been raised with such training can have considerable skill (figure 11). One participant told us how her sixteen-year-old son provided a meal for his family:

> One Sunday in March, he and a buddy launched their little shallow draft boat and set out to go fishing with rods, reels, and artificial bait. The fish weren't biting, and John caught only one bass. Since that would feed neither him, his fishing buddy, nor his family, he decided to turn that fish into a meal.
>
> He motored his boat to the local crab dam, banked the boat, and scavenged the area for abandoned strings. After finding several, he then used his pocket knife to cut the bass into several pieces and tied each piece to the end of a string. The strings were then tied to pilings and left to hang in the water. Shortly, blue

Figure 11. Richard's grandson Keith when he was about thirteen years old learning to field dress a rabbit he shot. Photo courtesy of Richard Borne and Haley Metzger.

crab took the bait, and John scooped them up using the fish-dip net from the boat. They put the blue crab on ice and brought them home for supper.

Another Sunday, John and the same buddy went fishing again—same boat, same tackle, same place. This time, they caught no fish at all, so they returned to the crab dam. Having nothing to use as bait, they used their cast net to catch bait. They caught a mullet and a croaker. Having dropped their pocket knife into the water earlier, they found a broken bottle and used it to cut the fish into pieces to bait the crab strings. This time, they netted about two dozen crabs.

John returned to his family's camp where his mother was entertaining company. He gathered the necessary items to boil the crab to perfection and then served them to her guests.

This story was a source of great pride. Her son had learned some skills and showed real resourcefulness. Although still in high school, he was able to provide a meal for an entire family (on more than one occasion) and could even improvise when he lost a key tool, like a pocketknife. Just as importantly, he learned to independently make some choices—like taking the initiative to boil up the crabs and serve them to her guests. This is the kind of self-reliance, coupled with generosity (preparing and sharing the crab with family and guests), skill, and deep family connection that coastal residents prize.

Students at Nicholls were very aware of hunting and harvesting as enculturation tools. We were impressed with the degree to which students, most of whom were not yet twenty, were already envisioning themselves as parents and sorting through the values they thought important to teach children. First, they were aware that self-provisioning activities were a type of knowledge or competency they were responsible for passing on. For example, one student identified the purpose of the family camp as a place to transfer knowledge. After supper, he said, the men sit and tell stories, often about hunting or shrimping. The children are the indirect audience. "The kids stand around their fathers with their soft drinks, listening to the stories, knowing one day that will be them with a beer in their hand." In the morning is when "the real teaching will begin." The student said, "Fathers dress their sons, show them how to hold the gun, and so on." The morning shoot is followed by more lessons. "Lunch is cooked by the adults as the kids watch and learn; one day that will be them standing over the grill showing their children."

As in the example above, students used their essays to imagine futures in which they would be the teachers for their own children. For example, a student who wrote about his love of duck hunting and training dogs said in class discussion that he needed to learn hunting skills because one day he needed to teach his own children. Another student said for him duck hunting is a special way for parents and children to bond. The duck hunter he interviewed said, "Some of the best memories I will have with my son will be out here, hunting side by side with him." A third student interviewed a deer hunter who had considered quitting, but "he said, having the family together is one of the main reasons he still hunts, and that it is how his children got to know their family." Clearly, he doesn't mean that his children don't know who their parents or siblings are. When this hunter says, "got to know their family," he is talking about knowing other people in a way that happens when you spend hours together—tromping through the woods, building a duck blind, or fishing side-by-side.

Beyond the specific skills or connection to family, the students were able to crystalize an idea that many other people in the region tried to explain to us: hunting and harvesting are the way you raise children who become good parents, active community members and citizens, and caring neighbors. This, in other words, is how you create good people. Cory, a freshman, wrote an essay in which he interviewed his grandparents, who stressed that they taught their children both an occupational skill and a lifestyle when they taught them to shrimp:

> Alton and Lou Ann have taught four of their seven children to shrimp. "We taught them so they can learn to do it on their own and be able to make a living from it," Alton said. They feel that if they teach them how to shrimp, they have fed their children for a lifetime. Shrimping was so embedded into their lifestyle that they felt the need to pass this knowledge to their kids. Teaching children how to shrimp is a very important aspect of the shrimping culture because it not only passes down the knowledge, but it continues that part of their lives.

Cory's older brother, Ryan, told him how learning to shrimp had shaped him:

> He learned not only to catch shrimp, but all the life lessons that came along with it. He remembers the hard work of pulling up the nets and separating the catch. His grandpa taught him how to provide for his family and the art of self-reliance. He remembers his grandpa always telling him how to live right. "He was teaching me how to be a man," Ryan said.

In this chapter, we have seen how people who engage in subsistence practices have goals beyond food, skills, or economics, even beyond the joy of the catch and the simple pleasure of sharing it. For many people, subsistence activities and harvested foods are ways to bind their families and raise their children to share their values and beliefs. People connect harvesting to culture or tradition in several ways. First, harvesting is a link to current family and community. Often, they are done collectively and the fruits of the labor are shared in ways that fill the senses—the quiet of the woods, the wonderful smells and tastes of delicious food. Secondly, people see harvesting behaviors and skills as connecting them to their families and communities through time. As older generations teach younger, even as teenagers teach preteens, people are aware of engaging with practices that have been transmitted again and again. Not only are their children learning skills and confidence, but these skills and activities are shaping their identity: personal identity, family identity, and a wider cultural and regional identity. Third, learning how to shrimp, hunt, fish, or garden and how to prepare and share food is integral to becoming a good person, someone with valuable skills that allow them to be both self-reliant and socially connected. This shared recognition of what makes a good person—a combination of self-reliance and deep community and family ties—is foundational to identity. Finally, people see harvesting activities as providing a mechanism for families to stay connected. They stay connected in order to learn harvesting skills, sometimes to harvest together, and, most importantly, such skills retained by a few allow for larger gatherings of extended family around highly valued and relatively inexpensive food. These large gatherings provide an important sense of belonging and well-being, in part because of the pleasures of delicious food. Feasting, simply put, binds friends and family together.

6

CAMPS, LEASES, AND CLUBS

Deer Hunting with the Brule Hunting Club

In 2012 and 2013, Mike Saunders tagged along on several hunts with Richard Borne, a resident of Labadieville, avid deer hunter, and recent retiree from the maintenance department at Nicholls State. Mike wound up meeting most of Richard's family and even camping out on the back of Richard's property. Below is a fairly long excerpt from Mike's field notes from the first time Richard invited him to the Brule Hunting Club,[1] offering an idea of what the camp is like, how it facilitates community, and showing how swamp deer hunts are conducted.

On Saturday January 19, 2013, I met Richard Borne in Labadieville, Louisiana, to join him as a guest for a weekend of organized deer hunts at the camp built and maintained by the Brule Hunting Club, of which he is a member. These clubs consist of groups of men (and various relatives) united by social or family bonds who form clubs with the aim of maintaining a communal hunting lease and associated hunting camp; members pay a small yearly fee ($250, in the case of this club) to pay for the lease and upkeep on the camp. The Brule Hunting Club was founded in 1950 and at present boasts approximately fifty members. The club maintains a camp (on the site of an old trapper's camp) about ten miles west of Thibodaux, Louisiana, on a canal that empties into Bayou Sherman, and leases a large tract from Williams Inc. (a landholding company) to hunt.

Richard, my host, has worked on the facilities staff at Nicholls State University in Thibodaux for nine years, taking a job there after retiring from a metal fabrication company. He has been married approximately forty-five years and has ten grandchildren and three great-grandchildren, all of whom he speaks of with great pride; he attempts to involve all of them as much as possible in the hunting culture in which he was raised.

After meeting Richard at 6:00 a.m. in Labadieville, I followed him a few miles to a boat ramp. We loaded our gear and guns into his 16-year-old grandson's small boat and headed out. As a teenager, Richard's grandson only owns a 14-foot johnboat (a steel or aluminum flat-bottomed boat) with a 9.9 hp motor. We had running lights on the boat, but otherwise only Richard's hand-held flashlight lit the way. After about 25 minutes on the water we arrived at the dock. A walkway of metal grate ran about thirty feet from the bulk-headed docks to the higher ground on which the camp sat; at least a foot of water was under the grate between them. There were pens behind the house that held a dozen or so hunting hounds.

As we entered the building I was surprised to see not only televisions tuned to local news shows, but a very well-equipped kitchen (with a large island and a commercial oven with a grill unit), two poker tables, and an alcove with a pool table. Moreover, there was a huge pile of homemade biscuits and an equally massive pan of breakfast sausage on the kitchen island; six or seven men stood or sat as they drank coffee, ate, and idly commented on the news.

Richard told me that at least a few of the older men stayed almost full-time at the camp during hunting season, acting as caretakers, cooks, and just enjoying the place and the company. Soon more men, often much younger, started appearing from the bunk room at the back of the building (where I would later sleep). I counted about forty bunks, plus six beds in a separate area with "Old Timers" written on the door. There were names written on the sides of most bunks and some were 'customized' with end tables, lamps, radios, and even small TVs.

After stashing my gear and returning to the common area, I found it was time to begin the hunt and headed outside to don waders and a safety vest and head for the boats at the docks, as did everyone else. At the docks we divided into groups and boarded boats that would take us to a drop-off point where an airboat would eventually arrive to take us deeper into the swamp.

The strategy behind an organized dog hunt is to block out an area of land with parallel lines of hunters; others are positioned at one end of these lines, thus hunters are situated so as to surround (on three sides) a roughly rectangular area. The dogs are then turned out at the end that remains free of hunters, with one hunter (the "driver") moving behind them (to encourage them with excited yelling and to keep them moving in more or less the right direction). The dogs subsequently work the area to spook any deer they find into moving—to be potentially shot by the hunter closest to the jumped deer.

Before long we heard the airboats start up back at camp. When the airboat arrived, we all boarded it (a full house with most of us sitting on the gunwales) for a trip up a very small cut through the swamp. We soon came upon a cypress tree with a small sign nailed to it identifying it as one of the stands. At the marked tree, we dropped off one hunter, a few hundred yards further we dropped another at a similarly marked cypress. My group (Richard, his grandson, and me) then unloaded at the next "stand." I now understood our stand to be a stand in the literal sense—we would be standing knee deep in water (and this was a high spot) while leaning against trees and awaiting any moving deer.

We were on a swampy fringe of Bayou Sherman, which drains into Grassy Lake (and then into Lake Palourde, then on to the Gulf of Mexico). We were now arrayed in two lines about a half-mile or more apart on both sides of the canal. Hunters were also positioned to block off the eastern end of these lines. The dogs were soon released—evident by the excited pack sounds as they headed out to search for scent and a trail. After this we simply listened for dogs approaching (baying if they were on a trail) to signal a deer may be moving through, as we watched for any movement that might indicate a "sneaker" deer was trying to move out of the area. Occasionally we heard the dogs baying. I was told they were probably either cold-trailing (on an old scent) or possibly chasing nutria.[2]

Soon Richard received a phone message from his son (a few stands beyond us) with an attached video of a doe walking past. However, the club had voted to take no does this season (though all members hold doe tags as part of their deer license). A recent epidemic of Blue Tongue Disease (spread by insects, it affects ruminants and prevents them from eating, thus is highly destructive to game populations) had caused a dramatic herd reduction, and the club hoped to replenish the herd through conserving does.

Eventually, although a bit early (the projected end to the hunt was 11:30 and it was only about 10:45), we heard one and then both airboats fire up. A few minutes later an airboat arrived with hunters from the stands beyond us already aboard; we boarded and got back to the boats and returned to the camp. At camp, a quick count among all hunters totaled two does spotted other than the one filmed (although at least two nutria were shot—I saw the tails, worth a $5 bounty).

The hunting is done by mid-day and then there is nothing to do but be social and relax. And a real sense of community was obvious, despite the fact that many of these men lived in different towns, worked at different jobs, and in many cases only saw one another during hunting season (as was evident from the greetings exchanged and the "catching up" going on between various hunters). There were, however, long-time ties between most of the men. I spoke with an oil field worker recently returned from a job in Nigeria who had only been home the last few months (more or less specifically for hunting season). He had been a member of the Brule club since 2007, joining after a previous club to which he belonged lost its lease. He noted that there were often marriages or other ties, beyond simply blood relations, that linked the members of the club. Moreover, he said some of them had grown up together. "That guy's uncle used to hunt with mine, so we knew each other way back." He noted that the sense of community and camaraderie weekends like this provided were as important to him as the hunting.

Later that afternoon I accompanied a few of the men on a boat trip to a nearby marina and bar and witnessed more of the temporary (yet lasting) community established during hunting season. After traveling up Bayou Sherman, through Bay Sherman and then into 4-Mile Bayou, a number of buildings came into view, dominated by a large marina with a roofed open-air dance floor. There were at least 100 people milling about or sitting and listening to a band (and a few older couples dancing). As we went ashore my companions quickly began exchanging greetings with people they knew, joining others who were visiting from their own camps.

Back at camp for dinner, I found the same general crew appeared to handle the cooking as well as the serving. Soon frog backs and catfish nuggets were carried around and offered to each person. These were all recently caught game. However, they didn't serve the frog legs. Those who caught the frogs saved those to eat at home. The catfish nuggets were from a few fish pulled from jug lines (single baited hooks attached to a float) that day—only a few for each person. They made an excellent appetizer before the very nice steak that followed. Soon after dinner, I was quickly asleep, sleeping better than I have in months.

When I awoke it was already 7:00. I went into the kitchen and found biscuits and a huge pile of bacon already prepared. This day's hunt would be along both sides of the low ridge of high ground behind the deer camp. Rather than boarding airboats, Richard, his grandson and I, and two other hunters simply walked across the swampy area behind the camp to reach the ridge and then stationed ourselves at the first four blinds[3] there. Beyond the ridge was the line of other hunters and beyond that swamp (all the way to the Gulf, noted Richard). Richard and I were in a tree stand about 15 feet in the air and the view was surprisingly different from that at ground level (where thick shrubs, palmetto, and other undergrowth block much of the view). Aside from a few brief comments, we primarily sat in the blind (a real luxury considering it was dry and actually had seats) and watched for "sneakers" while listening for the baying of the dogs.

We saw no deer during the hunt and returned to camp to find that only two does had been spotted by others. Soon after, everyone departed almost immediately. I barely had time to catch a few to say goodbye and thanks, although some of the old timers were in the kitchen doing the final cleaning. As had occurred several times that morning, when I thanked them I was told to "come back any time," as I could be a guest of the club now—all I needed to do was get in touch with one of the members or even "just show up."

✦ ✦ ✦

After exploring how subsistence practices work to pass on values and identity and provide opportunities for family and friends to bond and create strong community ties, we now look more closely at a set of structures that help people to maintain these practices. Specifically, we explore a common coastal strategy—having a hunting or fishing camp, joining a hunting or fishing club, or leasing the right to hunt or fish on property.

While pop culture images of subsistence activities in Louisiana can sometimes be celebrated as individual acts of self-reliance—think of images of isolated Cajun fishing shacks or the people spotlighted in television shows like *Swamp People*—in reality activities are just as often group-focused, institutional, or even corporate. While some people in our study prize hunting or harvesting for the opportunity for solitude and individual accomplishment, many others hunt or fish in groups and see harvesting as a time of socializing, like the men at the Brule Hunting Club. Sometimes a group is part of a nuclear family. More often, we saw multigenerational extended families or groups, sometimes linked by friendships acquired in school or work. In this chapter, when we examine the camps and clubs, we should see them as not only physical places where you harvest food, but also as a platform for subsistence-linked activities we have talked about: socializing and bonding, enculturating children, and a source for creating narratives about values, community, identity, and self. This chapter is also about another side of subsistence, gaining access to resources—lands to hunt, waters to fish, plots to garden. For many, it is clubs, camps, and leases that provide the infrastructure to make the socializing, bonding, and enculturating possible.

Clubs and Leases

The story Mike tells about his time at the Brule Hunting Club allows us to see hunting as very much a community or group-based activity. We can see that the members enjoy their before-hunt time with homemade biscuits and conversation, and the after-hunt time is all about visiting, making connections, and getting to know people. Even as people are more isolated during the hunt itself, they are part of a group activity, and each person has to understand their role in order to make the hunt a success (figures 12 and 13). Even when the hunt does not result in a kill, the time is not wasted. As Mike writes, "The hunting is done by mid-day and then there is nothing to do but be social and relax." These hunters have anywhere between eight and twelve hours of socializing after the Saturday hunt, and this social time is a key component of the function of such clubs. Mike wrote in his notes that one hunter "noted that the sense of community and camaraderie weekends like this provided were as important to him as the hunting."

In other field notes, Mike said that the hunters discussed everything from community news (or gossip) to political issues (particularly concern about

Figure 12. A member of the Brule Hunting Club is dragging an 8-point buck he killed out of the water. A deer that large can weigh between 160 and 230 pounds, and they have to be dragged, usually to a boat, to be transported back to camp and dressed. Photo courtesy of Richard Borne and Haley Metzger.

Figure 13. A lineup of the deer shot in a successful day of hunting for the club members. Behind the deer are the flat-bottomed airboats used to move in the swamp, where the water is shallow and a submerged propeller could get easily stuck. Photo courtesy of Richard Borne and Haley Metzger.

gun control) and conservation. For instance, the club voted to not shoot any does during the season that Mike documented. Richard told Shana that the decision had been controversial, particularly for "some of the younger members." However, strong social ties and a sense of respect for the knowledge of the older members allowed them to all eventually agree to not shoot does as long as blue tongue disease[4] threatened the deer population. The older members explained to the younger members that by taking such voluntary control hunters were showing themselves as true protectors of the wilderness and as a group of people responsible enough to not need state regulation. In other words, they defined themselves as the real conservationists.

Hunting clubs are grounded in a hard environmental fact: most people in south Louisiana do not have the option of simply walking to the closest woods to go hunting. Fishing is simpler because there are a multitude of free, accessible fishing spots, but the activity is still licensed and closely regulated. And even fishing is not permitted in some places. Hunting and fishing require access to where the animals are—land, water, marsh—and most land and many inland waterways (or canals) in south Louisiana are owned by individuals or corporations. In fact, some of the favorite fishing waterways are privately owned because they are canals that were originally dug by private industry or individuals. One researcher counted 661 named canals in coastal Louisiana noting that there are hundreds more without the word *canal* in the name (Davis 2010:161) and estimated there are about eight thousand miles of canals crisscrossing the coastal region (Davis 1973). Many of those canals remain owned by private interests. Outlets from lakes often have locked gates to prevent boats from entering, and companies sell leases to hunters or fishers giving them rights to enter those waterways to harvest game or fish.

Hunting brings the greatest problems of access. Two free options are to hunt on your own land (if you're fortunate) or to hunt on a state-owned Wildlife Management Area (WMA). In the coastal region where we worked, the largest is the Pointe-aux-Chênes Wildlife Management Area. Accessed from Island Road, the WMA is thirty-three thousand acres of marsh and wetlands managed by the Louisiana Department of Wildlife and Fisheries and set aside for hunting and fishing. Much of the area is marsh or water, so people need a boat to get access to prime hunting and fishing spots. The Pointe-aux-Chênes WMA has special hunting lotteries for hunters with physical challenges or mobility limitations as well as youth lotteries for deer hunting. People can and do use this and other WMAs. For instance, in his oral history, John Serigny mentions that he submitted his name for a state WMA lottery and was chosen to participate in a goose hunt in another part of the state.

Other people feel less favorably about WMAs. Paul Hingle, a resident of Port Sulphur, was interviewed in 2002 by Tom McGuire as part of a BOEM cooperative project to document the history of the offshore oil industry. He explains that if you don't lease land, you don't have any place to hunt because WMAs are not good options:

> As far as going on public land, I ain't hunting public land. It's too dangerous. You got too many people that don't know what they're doing. Them hunter education courses, to me, they're not, they're not teaching you know—I mean they try to teach you something in two days that you gotta learn about guns and stuff like that. You gotta learn that from the time you're a little kid growing up.... Well, if they had more public land it might not be bad, but you know there's a lotta people that don't want to pay the money.... The last couple of times I've hunted game reserves and all that was open to the public, you know, I've seen people hunting out there that was drinking. I've seen people to where they look like every tree you look behind there was somebody. I even had a friend of mine that was in a tree stand, jacked up a tree, and he heard some noise right at day—before it got daylight. And when he looked down like that there was a guy climbing the same tree he was in. So, where we hunt at, we got maps where everybody's gonna be, and I feel safer in the woods. (McGuire 2008:7)

Some people in our study, like Richard, own property where they hunt or fish. He often takes his children and grandchildren back on the property to hunt, with success. For example, in 2011 he shot a ten-point buck. Richard has said that even though his property is privately owned, he tries to be cooperative with neighbors who might be, for example, tracking a deer. He told a story about a time when he was building a deer blind, and some hunters came through tracking a deer who crossed into his property. He let them continue following the deer. After they got the deer, they came back and stopped to help. The blind, needless to say, got built a lot quicker. Richard told this story because he believes this spirit of cooperation is how it should be and because it represents his experience of the general attitude held by the region's hunters and property owners, although he knows there are exceptions. Richard explained that while people may be understanding enough to allow a hunter to continue a chase that began on other property, seldom would they just welcome a hunter to come onto posted private land without express permission.

A great deal of open, forested land, suitable for deer hunting, is owned by corporations. Corporations will often sell leases to hunt on their property, a tradition dating back to the early 1900s. At that time, the clubs that bought such leases catered only to very wealthy clients, usually businessmen from the North

(Davis 2010). These hunts, comprised almost exclusively of white men, were famous for their luxuries and servants. One of the oldest was established after the turn of the century by A. E. McIlhenny, famous for producing Tabasco hot sauce. Back in the 1920s, access to his club cost elite corporate executives the equivalent of more than $100,000 in today's dollars (Davis 2010). Certainly, elite clubs still exist. Community members have told us about a corporate deepwater fishing club that is basically a ship out in the Gulf where people are helicoptered in and are entertained lavishly. The stories tell of tycoons landing enormous fish by day and making powerhouse deals by night. One such camp became famous in 2004 when Supreme Court justice Antonin Scalia was accused of being biased because of a duck-hunting trip with then-vice president Dick Cheney to a floating camp owned by a Louisiana petroleum industry executive (Janofsky 2004). We did not meet people who had participated in those exclusive clubs, but we met lots of people who were members, or had been members, of some type of hunting or fishing club.

Today, clubs have become common among folks from all walks of life. A "club" can simply indicate a group of friends and neighbors who have organized to be able to afford access to a leased property. For example, one of our participants belongs to a group of several family members and friends who chip in together to purchase a duck-hunting lease from one of the larger corporate landholders in Terrebonne and Lafourche Parishes. Collectively they pay more than $5,000 a year to have access to land for duck hunting and fishing in some canals. They have formed a hunting club so that they can lease the land together legally. This situation is typical. Deer hunters we interviewed pay up to $1,000 a year or more to belong to a club, which gives them access to a certain number of acres and often the right to set up a camp or a stand. Sportsman websites are full of people advertising places in hunting clubs or seeking to buy into a lease or a club. A hunting lease gives the purchaser (or group of purchasers) the right to hunt for specific game on the property at specific times of the year. A property owner might lease his land out to one group of people for duck hunting and to an entirely different person or group for alligator season.

The reality is that land use is more restricted now compared to fifty years ago. And in coastal Louisiana, another problem that emerged in tandem is that there is simply less solid land. The changes in land mass are dramatic. As recently as the 1960s, coastal Louisiana was about 25 percent bigger (by approximately two thousand acres) than it was by 2010. Some of the most dire estimates (Marshall et al. 2014) hold that by 2067, almost all of southeastern Louisiana land that is not behind a levee will be gone (figure 14).

People we talked to often noted how little land is available for hunting and fishing compared to their childhood. Some land became inaccessible while we

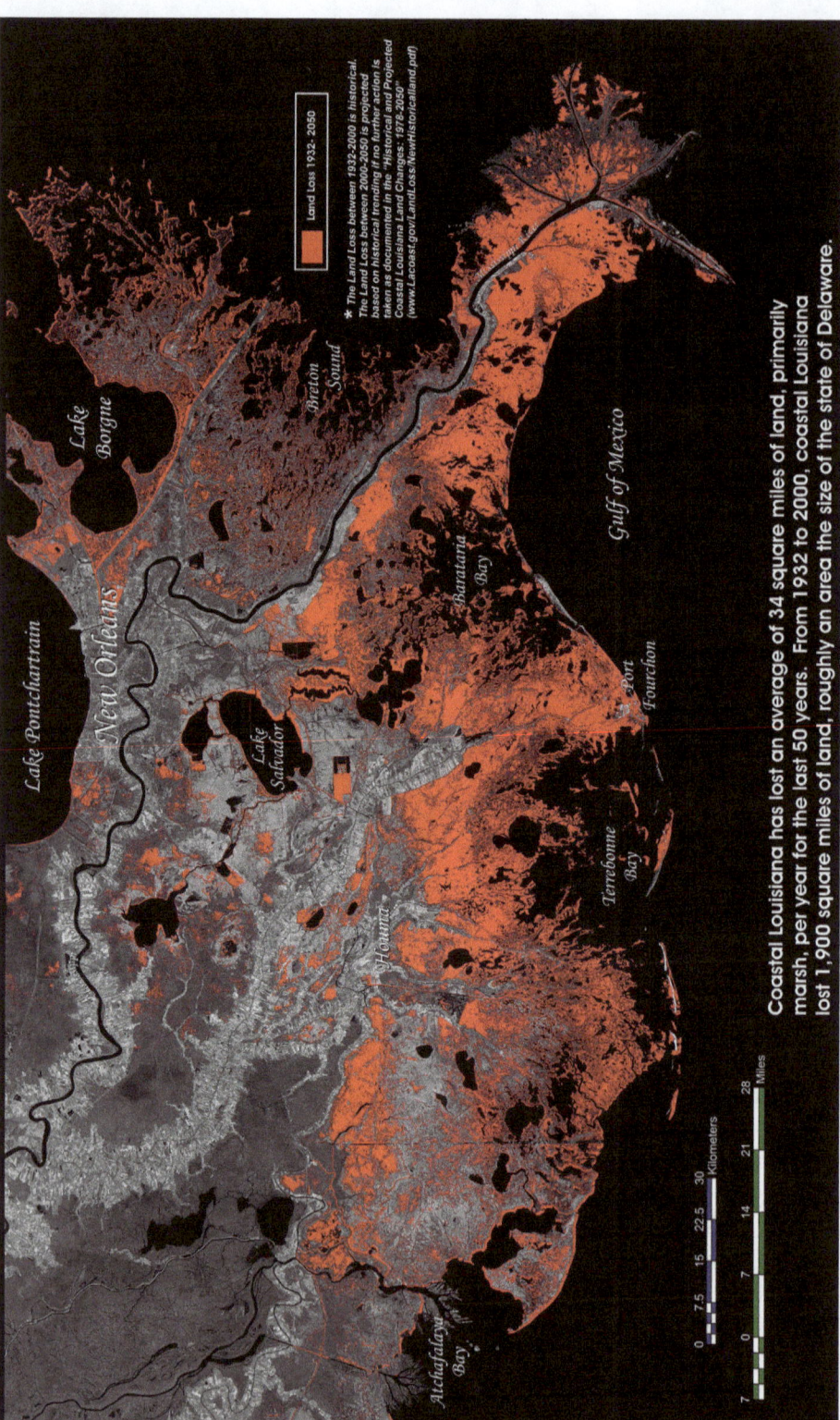

Figure 14. Louisiana coastal land loss from 1932–2000: 1,900 square miles of land. From "Southeast Louisiana Land Loss," a poster prepared by the US Geological Survey, the US Department of the Interior, and the National Wetlands Research Center. Public domain. www.LaCoast.gov/LandLoss.

were doing our fieldwork. For example, we talked with a seventy-five-year-old man who lived in Larose (Lafourche Parish) and had been an avid hunter and fisher all his life. In fact, he described himself as "a hunting fiend." And he noted how unusual such over-the-top enthusiasm was in the late 1950s and early 1960s because "most of the people, they weren't doing it because it was a pleasure. They needed the food." He, however, was "hooked on the hunting." But now, he says finding a place to hunt or fish is not as easy as in his childhood. He told us how Lafourche Parish has paid to build a levee around private land in exchange for a fishing landing. However, the parish access road has been moved, and the result is that the landing now is on private property. "So now they got a big sign out there that nobody can go. Now it's trespassing. So the only ones that can go is if you have a camp on the marshes in the back of the farm. If you've got a lease for hunting ducks, or whatever kind of lease you've got, a paper on your windshield, you can go. Now, the public can't go." In fact, there are a limited number of permits for the public; however, he argues that the permitted land is too restricted. He also pointed to a canal leading directly to a local lake that is now considered private property. "If you want to go to the lake, you're going to have to have a permit." There are other ways to the lake, but the routes are longer. "In my boat it takes me about an hour and a half to get over there. In an hour you burn a tank of gas. Nowadays that's about what? $40?" Combined with land loss caused by environmental change, shifts in access like this (which function effectively like privatization) mean that fishers and hunters face increasing barriers to access. The increased time and cost to get to a fishing ground can make the difference between filling your freezer and going without.

Hunting clubs are formed so that individuals can pool their money to have access to these leases. Or a hunting club can be formed by a corporate entity in order to develop a profitable recreational facility or simply to entertain or cultivate relationships with potential clients. Property owners will advertise their land as available to be leased. Hunting leases advertised in 2012 for access to between five hundred and twelve hundred acres cost $5,000 and $10,000 per year to hunt for deer or duck (or sometimes both). In one ad, the cost was $12,000 a year for seven hundred acres. Of course, the value of the land would vary (for example, by how close it was to waterways or whether the land was near known populations of deer). Once clubs are formed, they need to keep a certain number of members to keep individual prices down. Larger clubs will advertise for new members on hunting websites. We saw one hunting club advertising for a new member; they were leasing sixteen hundred acres and charging members $700 to hunt and an additional $150 to bring a camper onsite.

Although some clubs advertise for members from time to time, others' memberships are less open. People explained that people who hunt together need to know each other. Smaller clubs, like Richard's, usually find their members through family and community networks. And, although having more members lowers the price, too many people also can overuse the resource: people explained that clubs should not have too many members. Sometimes people have to wait for an opening to be a member in a club. There may also be other restrictions. Once a club has formed, memberships are often handed down in a family. One hunter mentioned that people have approached her about possibly "getting in on" her family's hunting club, but they were disappointed: admitting an additional member would not be possible. The smaller clubs often cannot allow additional members because of agreements with the owning land corporation about the number of people who have access or knowing which people have access.

Cost is one factor in these decisions surrounding club participation and hunting in general. For example, Richard says he always has a rough idea of how much deer he and his family have to kill (and how much meat needs to go in the freezer) in order to, as he puts it, "break even." In other words, to have enough meat in the freezer that the cost of buying the meat in the store would equal about the amount of the hunting club membership. While he has this balance in mind, he admits that he doesn't add in other costs to that calculation—rifles, ammunition, gasoline, miscellaneous hunting gear, or time spent building deer stands.

Because Richard and Shana both worked at Nicholls, they spent time each week talking about hunting, his family, and the club. However, when Shana pressed him on why his break-even cost calculation excluded what seemed to be a major part of his expenses, he just shrugged. "Well, I'm going to hunt." So, for Richard, the bottom line is that general hunting costs are just part of his life costs. The deer camp, however, needs to pay for itself in order for him to justify it. In his view, as long as he kills a deer and substitutes venison for store-bought beef, the membership is cost effective. The next chapter focuses more on the complicated economics of hunting and harvesting, including more discussion of people's investments in camps.

Camps: From Condos and Jacuzzis to "*Camp* Camps"

While the word *camp* in some areas of the US might imply a rough-hewn cabin or shack, down a dirt road, deep in the woods, this might or might not be the case in coastal Louisiana. What people call their "camp" can vary from something resembling a luxury condo all the way to what people call a "*camp*

camp"—meaning a much more basic shelter. Certainly, the camp used by the Brule Hunting Club is off the beaten path. Here is more detailed description from Mike about his trip to the hunting club:

> The ride out was fascinating, if a bit eerie (the moon was barely a sliver, and the night clear, which only added to the mood). Trees draped with Spanish moss hung over the canal, creating the effect of a tunnel through the darkness. With only the flashlight swinging back and forth across the water to light our way—variously illuminating the Spanish moss hanging from above, partially submerged trees, and floating logs and other vegetation—the ride took on an almost ghostly atmosphere, as if from a horror movie set in the swamps. It was about twenty-five minutes on the water (proceeding from the smaller canal onto a larger one) before we arrived. We had passed only one derelict camp thus far, but now a boat dock (or, more specifically, an amalgamation of docks) with several boats appeared.

But Mike was surprised when he stepped off the dock and into a building with a large, modern kitchen, a comfortable living room with multiple televisions on satellite service, and even a pool table. The Brule Hunting Club camp was built on top of what was originally a trapper's hut. They have enough land that a few younger members bring travel trailers that they park on the property to stay in during hunts. And all members have to own a boat because the camp is only accessible by water.

Camps are not only often this well-equipped, but also sometimes less isolated. Coastal Louisiana settlements are called "linear communities," that is, they are built along one main road that runs alongside the levee of a bayou. Marsh land is seldom suitable for supporting a building, so solid ground for building is at a premium the closer you get to water. In lower coastal communities, like Pointe-aux-Chênes, Dularge, and Dulac, many "camps" sit right across the road from similar looking single-family residences. For instance, community researcher Lora Ann Chaisson owns a house in Pointe-aux-Chênes that sits across the road from a "camp" owned by a couple who live in Belgium and visit twice a year. Truthfully, the "camp" is actually slightly larger than Lora Ann's house. Like Lora Ann's house, the camp is located on the same road as the local school and grocery store. Not off the beaten path, this camp is on the main drag.

Coastal communities like Pointe-aux-Chênes, Cocodrie, Dularge, Dulac, and Grand Isle are full of fishing camps. As with hunting clubs, the words "fishing camp" have a range of meanings. Some people described small trailers they have stuck onto a patch of land in a swamp or marsh, a houseboat docked to the shore, or they might have a small building, built over time by friends.

Others described four-bedroom raised houses with boat launches and a hot tub. For example, dozens of camps are crowded on the seven-mile stretch of Grand Isle. These are premium fishing camps, which in 2013 cost anywhere from $400,000 to more than $800,000 for three bedrooms with up-to-date kitchens and baths. One professor at Nicholls, whose family members are avid fishers, told us that, for her family, the ability to buy a camp at Grand Isle was a "lifelong dream." Even in a less crowded community, like Dulac, camps in 2013 cost between $80,000 and $200,000, depending on the amenities. In some areas the lots alone cost about $45,000. In other areas camps can be cheaper; in 2013 we found an online listing for a boat dock with a one-room, rough board shelter attached for $30,000. So, shelter spaces with a place to fish are available to a range of people.

Camps can be made more affordable when people band together: pooling money to make purchases, working together to slowly build rooms and expand over years, and handing them down to the next generation. Almost everybody that we talked to during this study had a camp or participated in going to camps. We met janitors and doctors, dentists and college professors, oil field workers, firefighters, local government officials, hairdressers, workers at the shrimp factory, students—all went to camps or had camps or had friends or family members with camps or used to have a camp. Often, they joined forces in order to be able to own the property.

For example, John Serigny's father and a group of friends purchased land in 1964. "This was my Uncle Gus's camp. He's the one who made my father purchase that, his camp." The camp had originally belonged to Gus's son-in-law, who was selling. "He didn't want strangers buying the camp. So he had my dad and a few of his friends buy the camp. Families are very close." Serigny, over time, bought the camp outright and has been hunting there most of his life. Serigny's duck camp in Leeville, representative of many family-based duck-hunting camps, is a simple one-room wooden building with a fishing pier outside and lined with bunk beds to accommodate a few friends and family. Originally the camp had three rooms, but the interior walls were knocked down to create one large open space with bunks. "There's a bathroom, but other than that everything is an open room. It's just a camp," Serigny says. Then he explains that it's a "camp camp."

Like Serigny, Felicia's investments took place over time. She slowly built her own fishing camp over many years, and her account underlines the importance of networks and relationships in creating and accessing camps. In her field report, researcher Tiffany Duet wrote about Felicia's childhood memories of trips to camps:

> Her family didn't own a camp in Grand Isle but stayed there with friends as often as they liked. Her father helped his friend install electrical wiring for his camp

trailer around 1965. With later improvements, a shower was added to the camp so that they didn't have to clean up outside with a hose.

While Felicia's family did not have their own camp, they had access to one. Later, Felicia, like others, was able to have a camp because she built her own largely by hand. She started when she was in her thirties and built it slowly over many years, starting with a boat she converted to a houseboat:

> She purchased a 55-foot aluminum hull and soon got to work. Although she paid for help with wiring and roof construction, she built the rest of the structure nearly single-handedly. On occasion, family and friends would perform small jobs in order to help her finish the boat before she was 50—and to win a bet that she would do so. She did, and with a few family and friends, they sailed the boat to its current location.

All of these approaches represent great amounts of money (cumulatively) and large investments in time, relationships, and networks. Without this, these ventures would not have been possible.

Camps can also be a business, a source of income for some people in coastal Louisiana. Team member Wendy Billiot bought a camp in Dularge, not too far from her house, and spent many days restoring it to rent it to out-of-town fishing customers. Another couple we spoke with work for a company that owns charter fishing boats as well as camps in Cocodrie. The husband acts as a fishing guide, and his wife offers visitors breakfast, packs them a lunch, and makes them her specialty—gumbo—for dinner. As he told us, "We call them *camps*, but it depends on your caliber of lifestyle as to what you call them." He and his wife reported that the out-of-towners who rented often called them cabins or summer houses. And most of them stayed only for long weekends.

The numbers of these camps for part-time residents are increasing throughout bayou communities:

> In 1979, there were 244 housing units in Terrebonne Parish that were classified as "for vacant seasonal and migratory use," according to the 1980 census. In 2005, there were 2,500 fishing camps on the tax rolls, consuming about 328 acres of parish land. That represents a 924 percent increase in twenty-six years. (Solet 2006:55)

Cocodrie is exhibit A for the conversion from a bayou town to a cluster of camps (figure 15). There are more houses now than fifty years ago, but only a small handful of people actually live in the former fishing village.

Figure 15. By 2020, the fishing town of Cocodrie had only one full-time family residing. All other houses had been converted to camps. The camps, raised sometimes eighteen feet to avoid flooding, crowd together on what solid land exists. Photo by Chris Usey.

Some of the newer camps for out-of-town residents are gated off. Solet (2006) reports that for the most part the out-of-town residents do not participate in the social and sharing networks of the full-time (year-round) resident community. Solet (2006) notes a growing division between community member housing and out-of-town camps, where out-of-towners build their "camps" among residential housing, rather than in swamps or backwoods areas and often have much nicer raised housing. In fact, the camps of out-of-town residents are more likely to survive storms and hurricanes than the full-time housing of community residents.

Clearly, hunting and fishing camps take a variety of forms, represent a range of investments, and are owned by people of all income groups. Camps serve as critical social institutions: these are places where children are socialized, colleagues bond, and community members share resources and talents. John Serigny's daughter held her wedding at the camp. One significant difference between the designer "condo camps" on the main thoroughfares and the camps most of our participants enjoy is the amount of work required to keep the camp functional and who performs that work. Camps located in the center of communities benefit from the same flood protection and amenities, such as levees and maintained roads, as the houses of full-time residents. They are also more accessible to craftsmen and paid professionals, should the owners want to hire someone to maintain or work on the property. Camps built outside these areas require additional upkeep. For example, at Richard Borne's camp, a fairly large place located on a spot of land surrounded by swamp, the ever-growing vegetation must be cut back and snakes cleared from the area, which happens on work days for club members during the off season. These less elaborate camps often flood and are subject to land degradation. Students report some of their earliest family work duties are participating in camp upkeep, clearing growth, painting, or shoring up wooden piers. With their sweat equity, the students often feel they have a stake in the camps in a way they do not have in their family homes.

Increased numbers of hurricanes are making family and community time investment even more important. After Hurricane Isaac in September 2012, John Serigny's camp was left mud-filled, molded, and with some structural damage, requiring cleanup:

> We're on our third one. The first camp, the old camp that we had, that my dad bought in 1964, we had until Katrina in '05. But then we built this camp [pointing to a picture]. It was 18 × 24. And for a one-room camp it was awesome. That camp we had from '06, after Katrina, until [Hurricane] Gustav in—what year was that? 2009? So, I mean it was a big disappointment to everyone.
>
> The year that we were building *this* camp, we did not duck hunt at all. All we did on weekends was go out to the camp and work on the camp.

"Looking for a Long-Term Relationship"

Networks are important in securing a wide range of hunting and fishing access, and camps, clubs, and leases play a role—as a thing you share and swap. For example, people often mention in interviews that they hunt or fish in northern Louisiana, Alabama, or Mississippi. Sometimes they pay into a lease in these places, but more frequently they are part of a swap. For example: you have a fishing camp in Louisiana near the coast. Your friend (maybe someone you met in college, in the oil field, or who married a cousin) has a deer camp in Mississippi. So, during deer season, you go to his camp and hunt on the land his club leases because most clubs will let you bring one guest.[5] At a later date, your friend will come to your fishing camp. This way, through reciprocal exchange, you get access to two camps for the price of one. These arrangements are common. For instance, Jerome told us that he has been invited to hunt in north Louisiana, Mississippi, and Alabama, but has not found time to get over to Alabama. Richard's son belongs to a club in north Louisiana. He and his dad swap hunts at their clubs. Joe Autin has swapped alligator hunting trips for deer and pheasant hunting in Kansas.

The internet has made such swaps easier. In some cases, the swaps involve pure barter, as in this Craigslist ad posted by a deer club member in Minnesota hoping to exchange access to leased land for shrimp. The ad included a photo of a man posed with a ten-point buck:

> **Whitetail bow hunt for Shrimp—$2500 (Minnesota)** I'm looking to TRADE an Oct. or November—5 day archery whitetail hunt for shrimp 400 lbs.
>
> Deer hunt Minnesota. We are located near Buffalo County Wisconsin and border northern Iowa! Our deer hunts produce! We specialize in guided whitetail deer hunts. We harvest 125" to 180" class bucks each year. Let us know if you are interested and I will be glad to get you more information. We have excellent deer leases for our clients. Trophy Bucks!
>
> Minnesota is a sleeper state!
>
> I need shrimp, prefer peeled and deveined, will consider tail & head on, prefer 16/20 count. I will need by August 15th . . . will have to figure out shipping logistics. I can use frozen. . . .
>
> Looking for long term relationship! (Craigslist.com, June 15, 2014)

This ad uses humor to underline a serious proposition. No, this isn't online dating, but there are similarities. These exchanges are transactions, to be sure, but they are also potentially personal and enduring—way more than a one-night stand, as this ad makes clear in its closing statement ("looking for long term relationship").

As we saw with Felicia's story, people also open their camps to friends. Felicia's family did not jointly own the camp she knew growing up, but her dad helped his friend with skilled labor (including all of the electrical wiring) and in exchange he was able to bring his family to Grand Isle.

Camps, Clubs, Leases, and Values

Despite the growth of designer camps on the main roads of bayou communities, camps of some sort are common among people in coastal Louisiana across income groups. Our research also found that clubs and leases involve people of all incomes. As we have seen throughout this chapter, these sites of subsistence practice serve multiple purposes, from socializing and skills training to food harvesting, preparation, and sharing. The camps, like subsistence activities, are not one thing. Camps and clubs can be collective work enterprises, individual DIY projects, family getaways, or even commercial businesses. In part because of pressing geographic, environmental, and corporate realities—including the continuing loss of access to fishing and hunting areas as well as actual loss of the land itself—camps, clubs, and leases have become tightly woven into bayou family and community culture. There is a more purely economic side to camps and clubs: memberships and leased land can easily result in enough harvested meat to fill a chest freezer. That counts. The costs and payoffs, however, cannot be reduced to dollars and cents because of the fundamentally complex nature of subsistence activities. As we examine in the next chapter, these economics are always linked to larger concerns of family, friends, community, personal identity, and values.

We made an effort to meet people who were interested in the widest range of hunting and harvesting possible, and quite a few were involved in multiple activities. We found more people involved in duck hunting than deer hunting and fewer still involved in activities such as dove, pheasant, or goose hunting. Although many people expressed an interest in deer hunting, financial realities block many people's participation. Deer hunting is something that people in a coastal region may move in and out of, as their networks shift. Deer, unlike ducks, require mostly solid land to live on, a habitat which is in short supply in coastal Louisiana.

Fishing costs can be highly variable. For example, rods and tackle can range from a set of simple equipment that is practically free (it can be borrowed or purchased used, rather than new) to sets of complex or sophisticated equipment that cost hundreds or even thousands of dollars. We spoke with people who had made significant investments in fishing, and others for whom fishing was a low-cost, out-of-the-backyard activity.

While there are deer-hunting clubs, like the one to which Richard belongs, in the upper parts of Terrebonne and Lafourche Parishes, hunting usually requires buying into a lease or waiting to get into a lottery on public land. And the clubs sometimes set limits on their members. Otherwise, men tend to go into Mississippi or north Louisiana to deer hunt. Duck hunting is more accessible for coastal people because the environment is richer for waterfowl and networks are usually closer to hand, and, in many cases, like John's, the camp or lease access was often set up in earlier generations.

7

"WORTH IT" AND OTHER MEASURES OF VALUE

Snapshot: Corn Picking in Chacahoula

When team member Tiffany Duet wants fresh tomatoes or corn, she calls Arthur Bergeron. We met Arthur in chapter 5, where you can see a picture of him as a child catching fish with his cousin. He lives in a small town in Terrebonne Parish called Chacahoula. Located between Houma and Thibodaux, Chacahoula was founded near a bayou. Arthur is a multitalented person who, when Shana visited him, showed her the shrimp boat he had just completely refurbished, including new welding. This snapshot draws from Tiffany's field notes and an oral history Shana recorded.

When June rolls around, lots of folks in Terrebonne and Lafourche Parishes start waiting for their phone call. The corn will be coming in at Arthur Bergeron's. Every year, Arthur plants a large garden and sells the produce, but he only sells through word of mouth. His wife keeps a list of people who have bought in the past, but they never put up a sign. Many years ago, they went to some farmer's markets, but it wasn't for them. Instead, every year they make the phone calls from the list. This way of selling seems to suit them best, where the practice retains a feel of being among friends and family but also pays a few bills. "I just want to get my money out of my hobby," says Arthur.

Arthur is a retired mechanic who lives in the same house his grandparents owned, although he has extensively rebuilt and renovated portions of it himself. The garden he plants is partially on the same land his grandfather planted, although he has expanded the space. He tills and harvests with a tractor his grandfather bought second hand in the 1940s, keeping it running with his skills as a mechanic. Arthur does not plant the same varieties of corn as his grandfather; he chooses to plant newer hybrid sweet corn varieties that his neighbors enjoy eating.

When Arthur was young, he was an avid fisherman and hunter. "When I was a kid, a teenager," he remembered, "our parents would drop us boys off on the levee [of the bayou] on Saturday morning and pick us up again on Sunday afternoon. We would spend the whole weekend hunting and fishing." The boys would pack a sandwich but would mostly rely on catching a fish to eat, he said. His memory is that they did not usually come back hungry. [See photo of Arthur and his cousin in chapter 5.]

He learned to garden from his grandparents. He remembers his grandfather, who was a country doctor, also having a corn patch to earn extra money.

In 2013, Arthur planted 16 rows of corn, each 300 feet long. He plants his corn two weeks apart, so that not all the corn comes ripe at the same time. Tiffany, a regular customer of the Bergerons, was at the 2013 picking. As she put it in her field notes, on the first day of corn picking, June 6, Arthur and crew made 43 sacks of corn, with 52 ears in each. The Bergerons sold about 30 of those sacks and gave away the rest to family and friends. Of those they kept, they boiled some to eat on the cob, and they froze many ears on the cob to use later in corn soups and seafood boils. His wife also cut some kernels off of the cobs to freeze for soups. Tiffany reported that about two days after the first picking, Arthur and his wife picked the "scraps," about 700 ears, which they gave away. A "scrap" is an ear of corn that has irregular kernels. Arthur doesn't sell these, but they make good meals for the people who get them. Tiffany reported that on the Bergerons' second day of corn picking, June 13, Arthur and crew made 48 sacks and sold 40 of them. The sacks sell for $20 each.

"That's how we fund our garden," Arthur explains. The money from the corn pays for tractor fuel, fertilizer, and seed—not only for the corn, but also the beans, the tomatoes, the cucumbers, and all his other crops, including his winter crops. His winter plantings include turnips, cauliflower, broccoli, radishes, beets, potatoes, and cabbage, all of which he raises from seed.

Arthur says gardening on the scale he likes, year-round, requires some investment of money. For example, he has invested in building shielded seed beds to begin raising more crops from seeds.

The other crops they don't sell, except for tomatoes. Folks around Chacahoula know they can drive over to the Bergerons and usually buy some tomatoes. In 2013, Arthur told Tiffany that by June, he had picked about 200 pounds of tomatoes, and his wife had canned about 100 pounds, many to give away to family and friends. In addition, she had made up 150 cucumbers into pickles, mostly to distribute to family and friends. They had also picked about 25 pounds of beans, freezing most of those to eat during the rest of the year, with some to give away to family. Their annual corn crop keeps their garden going, and their freezer full.

Figure 16. Arthur Bergeron in his cornfield in Chacahoula, Lafourche Parish, in 2013, driving a tractor his grandfather bought before World War II. Photo by Tiffany Duet.

✦ ✦ ✦

Arthur's story presents some challenges to any attempt to cleanly separate capitalist, market-driven activity from recreational activity and self-provisioning. Clearly, his garden is all three and more. By taking care of this large garden, he takes care of himself, his family, and his community (figure 16).

In an article published in *Ecological Economics*, Melissa Poe and colleagues examine similar practices in fisheries and call this an example of a "more-than-capitalist economic system" (Poe et al. 2015:242). They explain:

> Subsistence exists in tandem with, and at times counter-to, the market-capitalist logics of commercial fisheries, constituting what Gibson-Graham (2008) describe as "diverse economies." . . . Commercial fishing operators are engaged in making a living, but their "living" includes the values of social relationship and food and cultural traditions. (2015:248)

Their study set out to test the link between retention of subsistence portions of catch and the market. They theorized that if the primary motivation was to maximize earnings, then fishers would sell all they could when prices were high. After examining twenty-six fish-price relationships over twenty years, they found no correlation between the market price of the fish and the amount of subsistence catch retained, with the exception of one species. They concluded that subsistence behavior cannot be predicted by profit maximization; therefore, other motives must be considered. They observed that subsistence and share systems[1] serve to improve human well-being because in this way people are able to engage in high quality-of-life practices and support social networks. In addition, they argue that these practices, like Arthur's, serve to strengthen community food security and food knowledge systems, allowing people to have higher-value foods and weather difficult times.

In the preceding chapter, the discussion of camps and clubs often shifted into talk about cost and value. People like Richard would explain why their club membership was, in his words, "worth it." And this phrase, "worth it," was the most common used in weighing costs and values, almost always rejecting a strict dollar-and-cents way of assessing value.

In this chapter, we explore the parts of subsistence that can be seen as easily translated to dollars and cents—like the investment costs to be a hunter, fisher, or gardener; the cost and value of processing food to eat immediately or preserve; the cost and value of food consumed and exchanged; the costs and values of camps, gardens, or club memberships. We offer several ways of looking at these questions. We listen to stories, like Arthur's, to learn how people invest and assign value to their activities. We also include data that show what kinds of investments, production, and consumption we are talking about; for example, how much harvest and eating we recorded among people in Terrebonne and Lafourche. We start with a few brief snapshots offering some ways of seeing how people decide what investments and activities are "worth it" in their lives.

DIY Camp

Many people who participate in subsistence hold full-time jobs. For example, when Felicia, whom we met in the last chapter, was interviewed in 2013, she held a professional full-time job. We discussed how Felicia started her DIY camp slowly, first owning only a boat and gradually turning that into a houseboat—a low-cost do-it-yourself approach to building a "camp." Here we take a closer look at the costs of this camp and the value she extracts from it.

Her camp is a simple houseboat, mostly hand built by herself and others. She owns equipment specific for fishing, a mix of items she purchased or built for the purpose. She mainly fishes using homemade jug lines, like the ones Mike mentioned were used by the Brule Hunting Club members (figures 17 and 18).

Figure 17 and 18. Felicia using an electric knife to gut a catfish she caught with a homemade jug line and making the jug line using PCV pipe and a Styrofoam noodle as the float. Photos by Tiffany Duet.

These are floating lines, usually left overnight. Felicia used empty plastic milk containers as floats to sit on top of the water and mark the line. People often bait these lines in the evening and then check them the following morning. Although Felicia recycles her own milk containers to make the floats, she must buy the PVC pipe, Styrofoam noodles, hooks, and weights to complete the fishing lines. While this technology is relatively simple and low-cost, it requires personal labor to construct and maintain as well as an understanding of how to correctly build the lines. Jug lines are often torn or tangled and must be cut. Replacement and upkeep are regular tasks.

Some fishing technologies would be more time-intensive to construct herself, and Felicia purchases those. In all, she has purchased four boats: a canoe, a "party" barge (a large flat-bottomed vessel, the size of a large room, that can usually accommodate many people aboard), an aluminum hull (which she built into the houseboat), and an aluminum skiff (a flat-bottomed fishing boat). New or used, these require monetary transactions for manufactured goods. Over the years, she has also bought crab traps and fishing poles, electric knives for filleting, and freezers to store food.

Built on the base of the hull, her houseboat includes repurposed, recycled, and purchased materials. She used her own and volunteer labor for much of the construction and paid professionals to install electrical wiring and a bulkhead, and to construct the roof. In other words, the houseboat required investments of money, materials, personal time, paid time, and gifted time. In turn, those volunteers have received gifts of fish caught off the houseboat.

There is no doubt that her ability to fund the houseboat is supported by her wage job. The camp also pays back by providing, over time, a significant amount of food for her, her family, and friends. The camp itself also serves as a circulating good. Its owner is able to invite friends and family to spend time there, providing a venue for hospitality as well as a free vacation spot and access to wild, highly valued foods for guests. In turn, these guests will provide the owner with other kinds of gifts, such as free labor or foods grown in their gardens. Beyond tangible benefits, she also receives intangible ones, including recreation and pride of accomplishment. Some scholars might call it "social capital," a term popularized by sociologist Robert Putnam (1990), which he adapted from Pierre Bourdieu (1977). Finally, the camp is connected deeply to her personal identity, to her earliest memories of who she is as a person, to her family ties, and to her sense of place and belonging.

Some of the things the camp provides (like the fish), the costs of production (like the PVC pipe), or the exchanges of services (like the free help in working on the houseboat) could be easily converted into a dollar value. Other parts of this system (like the time spent learning to make a proper jug line, the pleasure of sharing time fishing and cooking with friends, or the value of the camp to a person's sense of identity or sense of well-being) would present a challenge to any system of economics.

Bragging Rights

In this next snapshot, we return to Jerome, introduced in the first chapter, a former commercial shrimper who was working offshore on an oil rig seven days on and seven days off. He liked commercial shrimping, but he now enjoys the seven-and-seven schedule that gives him more time to hunt, fish, and garden because he can provide food for his family. He is an avid hunter, fisher, and gardener, filling his freezer (and his relatives' freezers) and using the food for more than just a few meals a week. He also receives and exchanges food with friends, for things like fresh eggs. Jerome said that growing up, his grandparents had a garden big enough to feed their entire family, and he tries to carry on the tradition. For example, he makes sure to have enough tomatoes for his family and the neighbors.

Jerome sold his large shrimp boat after Hurricane Katrina but trawls both seasons and keeps his commercial license so that he can harvest more than the 100-pound limit placed on recreational shrimpers. Jerome taught himself how to make his own trawl nets and trawl boards. He enjoys making them and says it's a productive way to spend any free time. His family eats weekly from the frozen and dried shrimp he stores. He also shares a lot of his harvest at shrimp boils and with people in the community.

Jerome is a big fan of hunting, both waterfowl and deer, noting, "We eat duck like crazy." He hunts deer with friends on their leases in northern Louisiana and Mississippi, and he has been invited to hunt in Alabama but has yet to go. He tries to kill the limit every season (three does and three bucks). He processes all of the meat himself, butchering all of the cuts and making his own sausage. Because of this, he does not have to buy meat at the store, though he will purchase some pork chops from time to time.

While he is duck hunting or trawling, Jerome will also go "fish fishing," as he calls it. He prefers speckled trout, redfish, bass, or white perch (bream), but not catfish. Sometimes he keeps his catch, but often, he will trade his fish for crabs with his Vietnamese neighbors who are commercial crabbers. For small amounts, they make a straight trade; if he is asking for enough crabs for a party, he makes sure to also pay them some money.

Exchanges are an important part of Jerome's subsistence practices. He regularly engages in trades with neighbors, such as the crabbers. Our snapshot of Jerome in chapter 1 discussed how he participates in a bimonthly community dinner where the men bring a dish made from their harvests—shrimp, crab, or deer harvest or vegetables they have grown. These acts of communal sharing form the networks that make it possible for him to have an even wider range of harvested foods for his family: people might leave baskets of crabs under Jerome's porch or call him up and let him know there are extra figs on their trees if his family wants to make some preserves.

In this example, Jerome and his family have a mixed economy where they use his subsistence labor—hunting, shrimping, gardening, and exchange—for meals on a daily basis. Like other area residents, they are eating from their freezers, their gardens, the bayou, or the Gulf, almost every day. Their equipment, including the shrimp boat, fishing gear, and guns, is constantly in use. Nevertheless, Jerome is weighing the cost of those investments:

> When I was young we always had a pirogue, you'd have to go buy a pirogue for us to hunt. Now, I've got a mud boat. Instead of me paddling the pirogue, I crank the motor up and go. But the cost is a lot different.

Harvesting, in other words, costs more today in part because mechanical equipment—like boat motors—need maintenance and fuel. But the value generated is also greater than the calories (or nutritional value) of the harvest. For example, Jerome said his favorite part of hunting is "bragging rights":

> I know one morning I went hunting, it was the opening of the big season and I went back there, I put my decoys out in the water. I shot twice. I went and picked my ducks up, picked my decoys up, and come right back home. Within about 15 minutes I was finished. I had the limit. I come in. I passed by a few of them, and they said, "Where you going?" I said, "I've got to go. I'm finished." He said, "You're finished?" I said, "Yeah, I got my six. I'm going home." That's about right.

He bagged his limit in fifteen minutes on opening day with only two shots. How much is that story worth?

He also values the objects he uses to hunt and fish, in part for the memories they contain: he still has the first rod and reel he owned when he was ten years old, the first cast net he made for his daughter, the first boat he owned as a teenager, and the first gun ever given to him, a gift from his godfather. In fact, he has kept every single gun he has ever owned. He has stories about all of them: stories about the fishing tackle from his childhood, stories about the toy boat he made for his daughter to play with during long hours aboard the shrimp boat when the family was trawling. In this case, even when the food is an essential part of the diet, hunting and harvest is also about other things we value—the stories, the skills, and the pride.

Mixing Business and Family

Our third snapshot is of Wendy, a member of our research team. Her profile illustrates that commercial and recreational interests and subsistence activities do not always have clear dividing lines. Wendy makes her income working as a fishing guide for sport fishermen as well as conducting swamp tours. She is also an avid fisherwoman, filleting the fish herself and using the catch to provide fish for her family freezer. Over the years, she fished and hunted with her four sons until they were old enough to go on their own. At the time of the project, her husband was ill and no longer hunting, but her youngest son, a teenager, was still at home, old enough to hunt alone. He was hunting duck and rabbits in the winter, fishing and crabbing the rest of the year. He and Wendy would go frog gigging (frog hunting) together.

Wendy also gathers other wild resources. For example, during our study, she discovered a wild blackberry patch. Not wanting the berries to go to waste, she made several trips to the patch. The result: six jars of blackberry preserves, six jars of blackberry jam, six jars of blackberry filling for cobbler, and two batches of blackberry cordial. She also made blackberry dumplings twice, blackberry cobbler once, put nine quarts of blackberries in the freezer, and gave one quart away. She gardens and regularly shares with her neighbors, who also share with her.

Wendy regularly provides fresh fish fillets for two older women who do not fish. Their husbands, who once provided fresh fish for them, have died. She reported that the women were always happy to see her coming. Sometimes, if she catches a fish she cannot use in her own freezer or that she does not have time to clean, she will keep the fish on ice and then later drive down to a local fishing area looking for families fishing alongside the road. Once, Shana met Wendy to collect some field notes and got a surprise—Wendy brought her an ice chest full of fresh oysters she and her son had harvested.

Although women actively participate in subsistence, not all participate as widely as Wendy, and so, as part of her business, she teaches women outdoor skills. For instance, she organizes fishing trips specifically for women who want to learn sport fishing. She also fishes to bond with her family. She told us this story about fishing with her cousins:

We were sport fishing for fish they could clean and take with them to eat later on. We caught red drum, speckled trout, channel catfish, lady fish, largemouth bass, and a two and a half foot black-tipped shark. Talk about variety!

On the way back from our fishing trip, we checked two crab traps I had in the water. We caught about a dozen and a half nice blue crab. We went back to the camp, and my cousin and I cleaned the red drum and the speckled trout for him and his wife to take back with them. After that, he boiled the crabs, and we ate them for our lunch. They were delicious.

Like Jerome, Wendy is a person whose family is eating from the freezer, garden, and Gulf almost daily. She works in the fishing industry, which provides income and some food. She also depends on noncommercial fishing and hunting activities to feed her family. In addition to getting what she needs (what her household can consume), Wendy likes to have a surplus of valued foods so that she can share food around and receive gifts in return.

We hope these snapshots offer some idea of the complexity of sorting out the economics and challenge any simplistic mapping of dollar values onto subsistence activities. Quantifying the value of the foods themselves, much less the experience of harvesting and sharing them with others, is a complex task. In the next sections we will review what we learned about the amounts and types of food people are consuming, exchanging, sharing, and, in some cases, like Arthur, selling. Then, we look more closely at the dollar costs of participating in subsistence activities, including another look at clubs, camps, and leases.

Production and Consumption: Learning from Food Logs

In south Louisiana subsistence happens year-round. In their food logs, participants recorded harvesting vegetables or fruits every single month of the year, although August and September are the slowest months. And while "seasons" in which particular species can be hunted or harvested are set by the Louisiana Department of Wildlife and Fisheries, game and fish are so varied and plentiful that people can, and do, harvest animal protein every month. Winter is for deer and waterfowl, spring and fall are for shrimping, and people fish and crab year-round. And when people do not have fresh harvested food available, they eat from their freezers.

At the end of the four-year project, we had collected food logs from twenty-six people who recorded a total of 2,458 entries for foods eaten, harvested, preserved, or shared. Some of these entries were people simply recording a food they harvested ("my son went fishing"); other entries were for shared

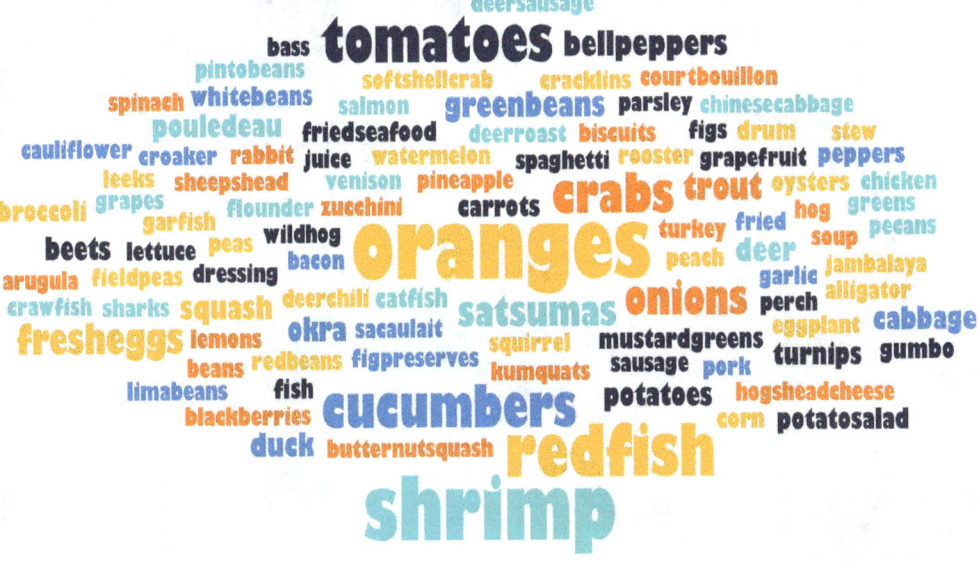

Figure 19. Word cloud showing the relative prominence of harvested foods reported in our logs. Items that were two words (like "poule d'eau") are written as one word ("pouledeau"). This shows that oranges, shrimp, and redfish were the most frequent food items that people caught, shared, and ate. Created using MAXQDA.

foods. Often people recorded the food as an ingredient in a prepared dish, like gumbo. People recorded a staggering 168 different types of game, seafood, fruit, vegetables, or nuts that were hunted or harvested, with some making the list frequently and others only rarely. One way to get an overview of the variety and importance of harvested and hunted food is a simple word cloud (figure 19).[2] Despite the use of multiple words for the same food, the cloud accurately reflects the relative prominence and abundance of specific harvested and hunted foods, like oranges and satsumas. The top ten most frequently mentioned subsistence foods, in descending order: oranges, shrimp, redfish, tomatoes, cucumbers, crabs, eggs, satsumas, green beans, and green peppers. These ten items alone are mentioned 836 times. Deer (sometimes listed as venison) was a close number eleven. Oranges and satsumas, common trees in many yards in coastal Louisiana, accounted for almost one-fourth of all foods shared and harvested in the logs we collected. If you add in shrimp, those three account for a little more than one-third of all food listed.

Significantly, the frequency of a food in this list doesn't necessarily indicate its value. For example, crabs and deer meat are highly valued. But the sheer abundance of citrus—and the way multiple trees in your yard may ripen all at

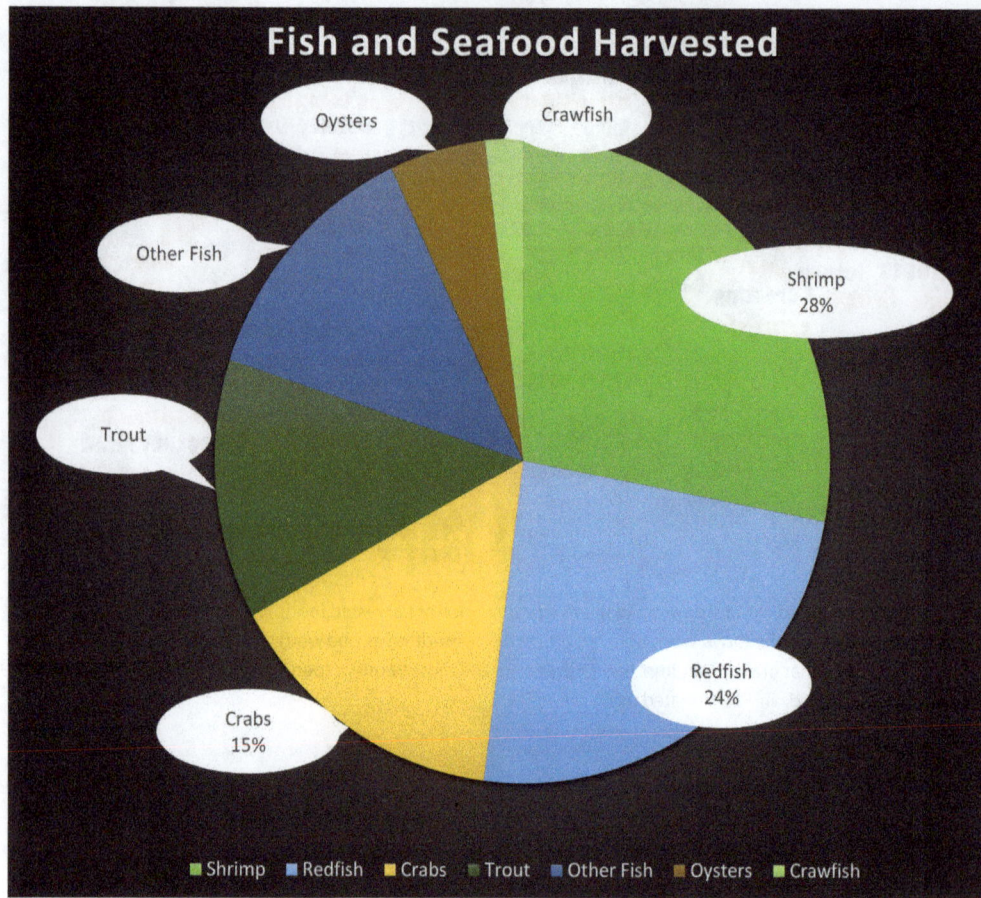

Figure 20. Fish and seafood harvested. Percentage of each type mentioned in logs over the course of the study.

once—make oranges and satsumas a heavily circulated, frequently mentioned item. Between November and January, bayou region offices frequently have a box of citrus in the breakroom with a sign saying "help yourself."

We divided the shared and harvested foods recorded into three categories: seafood and fish; wild game and waterfowl; and vegetables, fruits, and nuts. Three charts show the breakdowns by percentage.

Some of these entries were for harvesting; but almost as many were for shared seafood or fish (figure 20). Shrimp was the item shared most frequently. These entries were some of the most common cooked dishes mentioned, including shrimp gumbo and courtbouillon.[3] Fruits, vegetables, and nuts (figure 21) were the most frequently cited category of food harvested and shared. There were more than fifty total fruits, vegetable, and nut or fruit-based or vegetable-

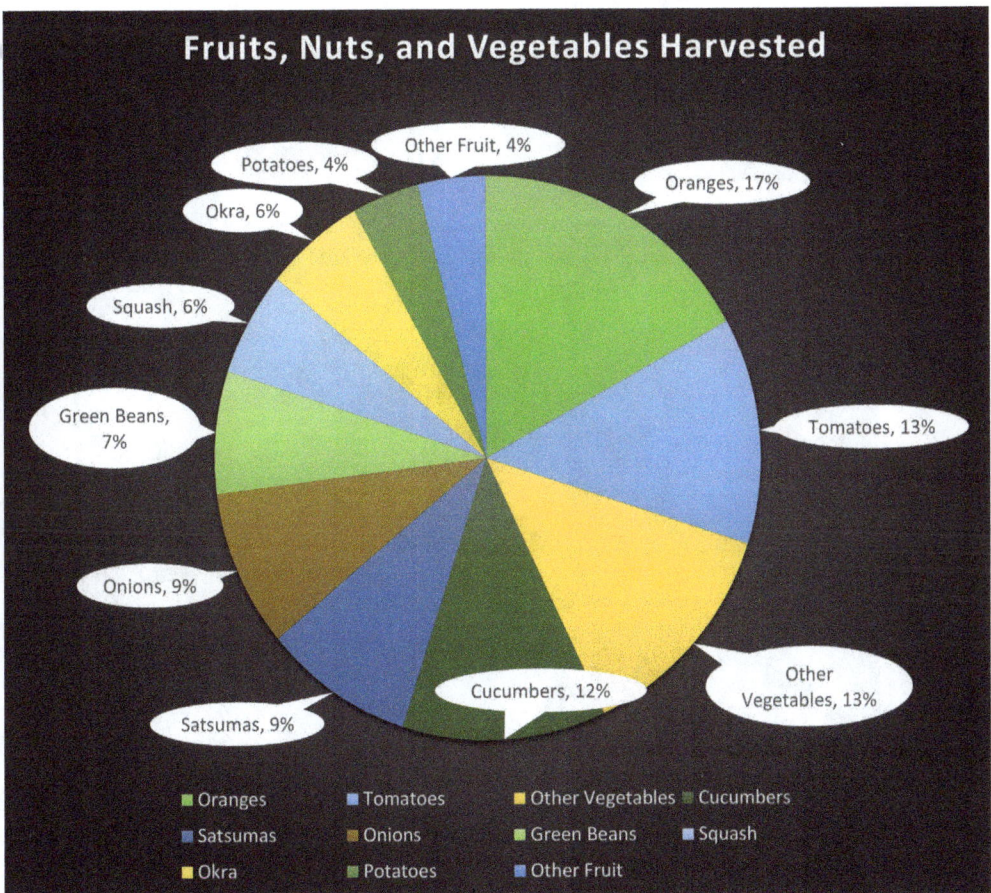

Figure 21. Fruits, nuts, and vegetables harvested or shared on food logs.

based prepared items mentioned (for instance, grape jelly or tomato sauce). Wild game (figure 22) was mentioned less frequently. For example, only about 5 percent of harvested foods logged were wild game. More often, game was shared in prepared foods, combining multiple ingredients, such as venison sausage or duck gumbo.

To get an idea of both how commonplace and how varied subsistence consumption is, we compare a week of food logs from three people, all from a similar time of year:

> **"Squeeze grapefruit daily":** First, we have a list of the harvested foods consumed or shared in late February 2013, by a husband and wife. They live in Terrebonne Parish. He hunts and fishes, she prepares most of the food, and they both garden.

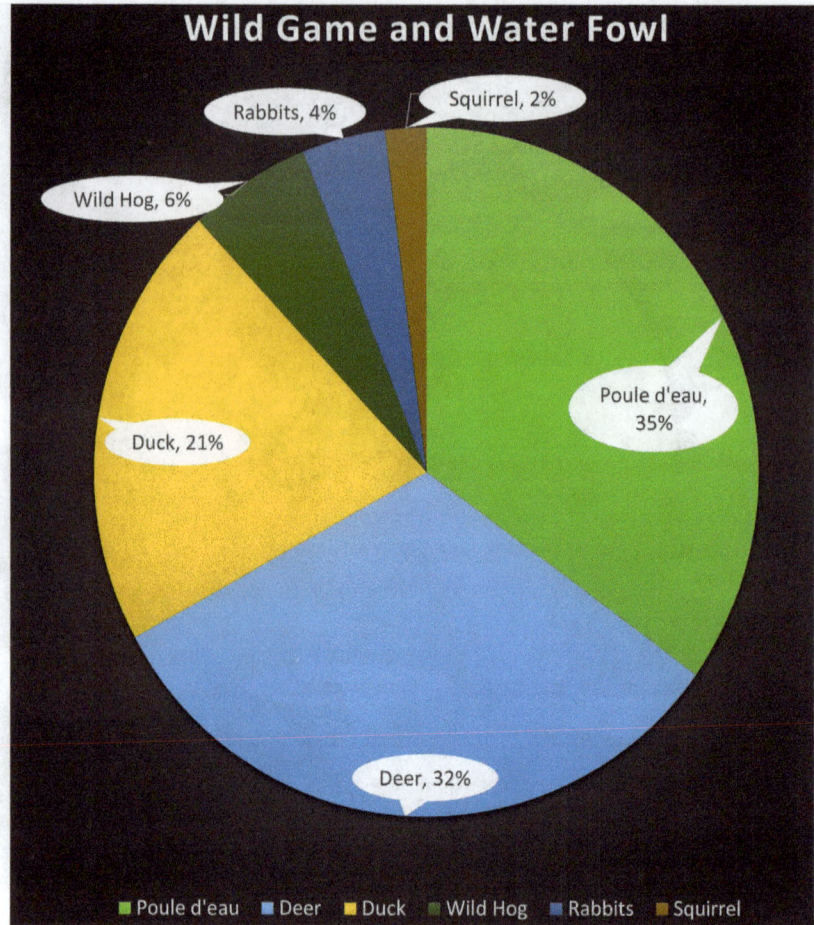

Figure 22. Wild game harvested. Percentage of types mentioned in logs over the course of the study.

Most of the foods were harvested from the husband's own garden; some were shared by relatives, friends, or neighbors. The husband recorded the log:

- **Eggs** for breakfast daily
- **Grapefruit**—squeeze for juice daily
- **Asparagus**—picking daily, eating daily as part of lunch or dinner (also shared with friends)
- **Speckled trout**—Wednesday a friend came by and dropped off a speckled trout, which they had for dinner
- **Green beans**—part of lunch or dinner most days

- **Honey**—extracted 15 pounds (sold 3 lbs to friends) and used some daily in tea and lemonade
- **Lemons**—used to make lemonade daily—drank every afternoon
- **Blood oranges**—given a batch by a friend—ate one but saved most to make wine
- **Tomatoes**—picked and ate daily with lunch and dinner

Cabbage, beets, and satsumas: We revisit Mrs. Dupré, whose story of oyster spaghetti opens this book. She wrote a narrative log of a week of foods from January 2012. Items harvested or hunted have been bolded:

Jan. 23: Today we had leftover smothered **cabbage** and yogurt I made. Harvested **eggs, cabbage, satsumas**. Homemade **satsuma** juice. Also gave 1 **cabbage**, 8 **beets**, 12 **carrots** to friend on this day.

Jan. 24: Today I made **cabbage** casserole. I also harvested **cabbage, beets, satsumas**. Made homemade **satsuma** juice. Ate **venison** sausage from hunting, and **arugula** from my sister.

Jan. 25: Today I made **broccoli** soup with things from garden. And we ate some **soy beans** out of our freezer that we grew. We also had homemade **satsuma** juice. Harvested **broccoli, satsumas, turnips, snow peas** from our garden.

Jan. 26: Harvested **cauliflower** and **satsumas** this day. Homemade **satsuma** juice.

Jan. 27: Today I fried fresh **fish** from my sons. Also harvested **cauliflower** and **satsumas**. Homemade **satsuma** juice. Also gave 1 bag of **kumquats** to neighbor and also gave 5-gallon bucket of **oranges** to neighbor on this day.

Jan. 28: Harvested **eggs, satsumas**. Homemade **orange** juice. Made **venison** burgers from **deer** hunting (son).

Jan. 29: Harvested **satsumas**, made **satsuma** juice. Cooked harvested **brussel sprouts** from the freezer, made a **venison** roast from hunting. Shared a **cabbage** with my daughter.

In all, Wendy collected almost a year of food logs from the Dupré family. During that time, the family augmented their diet from either their garden, fishing, or hunting for more than 90 percent of meals. For the total 173 days of logs, at least some portion of one meal came from gardening, fishing, or hunting on 158 days. In some months (May, June, and January), they ate from their garden or freezers every day.

They did not, of course, grow, hunt, or catch all of the food themselves. The Duprés primarily garden. Their sharing logs show that foods like venison, trout, teal, crawfish, and satsumas came in to the household from neighbors, siblings, children, and grandchildren. In return, they shared vegetables and fruits, both raw foodstuffs (like tomatoes, cabbage, and green beans) and processed foods (like soups, casseroles, and homemade grape juice). Over the course of seven months, from December 2011 through June 2012, they shared some food item every week. In May and June, the sharing logs record that food came in and flowed out on a daily basis:

> **Birthday parties and boiled crabs:** This log, also from January 2012, represents almost a week of meals from a tribal elder from the United Houma Nation who lives on Bayou Terrebonne. The week his daughter recorded the log also happened to coincide with his birthday:
>
> Jan. 16: Ate an Indian taco and leftover birthday cake. (He had caught 3 **redfish** and 1 **black drum** the week before and fried the fish and served it with white beans, shared with daughter and grandkids for a birthday party).
> Jan. 17: Had a second birthday party. Boiled **crabs** for a crab boil with 22 people this day. Bought the crabs from a local fisherman.
> Jan. 18: Fried **shrimp** from the freezer and cooked red beans from the store this day.
> Jan. 19: Caught 4 **redfish** this day, shared some with another fisherman. Also had dinner with granddaughter this day. She made turkey burgers and salad.
> Jan. 20: Caught 1 **speckled trout** and 1 **redfish** this day. Fried the **fish** and ate it with white beans [and rice].
> Jan. 21: Gave some **shrimp** to daughter, grandkids, niece and a friend this day. Cooked **shrimp jambalaya** from some shrimp given to him by his cousin at an earlier date.
> Jan. 22: Fried **fresh water fish** from the freezer.

These three sets of logs are from people who live on different bayous; however, they are all within a forty-five-minute drive of each other and their logs show their harvest in the months of January or February. And yet, the overlaps between the logs are only eggs, fish, and citrus. Two of the logs come from people focused on gardening, but they aren't planting the same things. One garden is focused on asparagus and green beans while still harvesting late citrus—lemons and grapefruit. Another garden is full of broccoli, snow peas, carrots, and cabbage. The citrus trees there are satsumas, oranges, and kumquats. Only the Dupré log mentions game because their son has killed a deer, resulting in meals of venison burgers and venison roast. The log from the Native American elder is the most distinctive because his diet is heavily reliant on subsistence protein (fish he caught, locally bought crabs, and shrimp his cousin caught). In addition, his log also makes it clear that a great percentage of his food is coming from non-self-provisioned sources, like white beans, red beans, turkey burgers, and Indian tacos (Indian fry bread usually topped with some combination of ground beef, chili, lettuce, tomatoes, and cheese). One striking similarity is that all three logs mention sharing food. In fact, over five recorded days, all three mention shared foods at least twice.

Good for You: Another Measure of Value

We can view the food logs as showing a type of economic benefit, replacing the cost of buying calories to eat. However, what we also see in the logs is that harvested and hunted foods are what might generally be called healthful—lean, low-fat, rich in micro-nutrients, low in additives, and free from the additional ingredients found in processed foods. Access to fresh, healthy foods is significant in terms of both economics and health. Anthropologist Katherine Browne, writing about the food traditions of an African American extended family in St. Bernard Parish, writes:

> In fact, food traditions that feature seafood are not only seeped in cultural meaning. Logic suggests they might also promote physical health and a sense of well-being. Research documents, for example, how fish in the diet produces serotonin, the "feel good" hormone in the brain that makes it easier to adapt well to stress. Moreover, fish and shellfish are the two best sources of an omega-3 fatty acid needed by the brain, called docosahexaenoic acid (DHA). Without adequate amounts of DHA, scientists have found, the prevalence of depression increases. Perhaps reliance on seafood had always helped the family to cope. (Browne 2015:49)

Browne's study documents the power of seafood boils and gumbos that draw together extended family gatherings as a way of finding comfort and rebuilding strength among members of a bayou community displaced after Hurricane Katrina (see Fisher 2017). But the practices she documents extend to how bayou residents practice self-reliance and care for each other in many situations.

The population of Louisiana on the whole faces particular health challenges because of high rates of poverty and nutritionally linked diseases. Chauvin, Dulac, Pointe-aux-Chênes, Theriot, Dularge, Isle de Jean Charles, and other small communities would all qualify as food deserts, because you have to drive to Houma (more than ten miles) in order to access a large supermarket selling fresh vegetables. Our surveys of the local shops showed that there was no local seafood available in down-the-bayou stores, very few fresh vegetables (usually only "seasoning" vegetables like onions, bell peppers, and celery) and potatoes. The situation gets worse as the stores get smaller the farther down the bayou you travel. That is, prices are higher, and there are fewer fresh foods, more processed foods. Given the food deserts that exist in many down-the-bayou communities, subsistence foods may play a critical role in people's diets. Significantly the concept of food deserts, which was created to highlight inequality in access to fresh, affordable, healthy foods, can sometimes have the effect of representing communities by what they don't have, by what they lack, or their deficits (Reese 2019:44–47). Our discussion of subsistence practices demonstrates how area residents practice self-reliance in spite of—or in the face of—these constraints.

Sharing: A Desire for Crabs

Sharing is a significant part of subsistence practice in coastal Louisiana. The three logs featured above are typical of all the logs we recorded. In all, participants recorded 844 distinct sharing events. They often mentioned sharing related to a food they were eating. Sharing likely occurred more frequently, as we have reason to think our data are underreporting both the circulation and sharing of food. Sharing is clearly a frequent and regular part of household culinary practice. Charles, a retired shrimper, explained how important sharing is in his community and in his way of life:

> The sharing of food is not necessarily something that you do on an exchange. It's something that you do because there's an abundant amount of seafood or something like that. If I go and I don't feel like cleaning my fish today, I'm stopping by your house and you can have it. And when I was shrimping with my father, we used to have my great aunt who lived by herself, barely making it on the money that she was receiving from the government. And every time we passed in front of her house coming home from shrimping, we stopped and put something at her house. Every time. Never expecting anything in return. It was just, you take care of the elders.
>
> And my father-in-law called me and tells me that the lady next door has pears falling out of her trees and she don't want them, go get them. My father, he eats Japan plums.[4] I'll meet somebody in the grocery store. "Tell your daddy to come get the Japan plums out of my tree." It's kind of like, whoever has an abundant amount of something and doesn't want to mess with it, you just give it away.
> ... So, for me to look back at some of the things that we take for granted—that somebody will stop by your house and say "Look I've got an ice chest full of fish" and just leave them for you, under your house.[5] Or call you up and say "Look I've put a box of crabs underneath your house."

As this account from Charles makes clear, many sharing activities are portrayed as "no big deal." They are an ordinary part of social life in coastal communities. Even as people told us about them, they often understated their importance.

Sharing is widespread but it's not random. The information gleaned from the logs and from our conversations suggests people are circulating food within networks, which include family members, neighbors, friends, coworkers, and fellow church members. Mapping out these networks systematically would offer added insight to how these relationships function. Although that detail was beyond the scope of our project, we can say that some people seem to be key circulators within the exchange networks. That is, some people take on a larger role in acquiring raw foodstuff, like garden vegetables or shrimp, and either recirculate it to others, or process it into cooked or preserved food before sharing it. Often these people are in linked occupations (like shrimpers or people like Wendy), they have schedules with some autonomy, significant time at home (like working seven-on and seven-off on an oil rig), or they are retired.

Retired people can often give significant aid to their children who are working full-time or demanding wage jobs, allowing the children to remain connected to a subsistence lifestyle by supplying them with food items, including fresh fish, boiled seafood, vegetables from the garden, and homemade preserves and pickles. The role of older relatives first came to our attention in the Nicholls student essays, which regularly cited grandparents as their connection to subsistence. For example, Tyler Sothern wrote about Sundays at his grandparents, where people gather weekly to eat and play cards:

> The meals at my grandparents' house often vary, but on this day we are having a very familiar dish for our particular family and for South Louisiana in general. The meal, consisting of fried shrimp and homemade French fries, is made possible due to my grandfather's connections here in South Louisiana. As the games start winding down and people begin to make their way out the door to resume their usually busy Sunday afternoons, everyone stops by my grandpa's freezer. Even after a large lunch consisting of shrimp, the freezer is still almost at the point of overflowing with bags of cleaned and deveined shrimp. The families take as much as they want to prepare for the week and make their way home, after saying the usual goodbyes to everyone. It may seem strange to people of other cultures that someone would store this much food to just give it away, but this is a picture of a usual Sunday afternoon.

The specialized role of elders sharing, organizing family gatherings, and maintaining social networks also routinely appeared in other forms of data collection. As with the retired couple fishing on Island Road (mentioned in chapter 5), their fishing not only supplies their supper several times a week, but provides a foundation for treating the extended family to a fish fry. Elders can be on the receiving end of sharing as well.

In fact, elders feature prominently among those receiving shared food. For example, Wendy often shares food with older people in her community. In his oral history, Glynn, the alligator hunter and shrimper we met in chapter 1, discusses his sharing with elders:

> I give it to the elderly people around here. A lot of people that I lease land with or use their property, I'll bring them stuff. Just people in general who, one day we may just be having a conversation in the store or something and say—like just now, a lady was telling me how much she likes soft-shelled crab. Well if I go shrimping and I catch soft-shelled crab I'm going to bring that lady soft-shelled crab. I know she desires it. She don't get much of it, and she would like to have some. She's a lady about her sixties I've been friends with all my life. Just gave me this hint that she likes soft-shelled crabs. So I'm going to bring her some. It's things like that. My family, my boys, they get their share of shrimp, crab, fish, ducks, deer, whatever. They'll come and use the property, and they'll hunt. But if we get extra we'll give it to them.

In Glynn's sharing network, as described here, are (1) people with whom he has relationships of reciprocity, including people he leases land with or who let him use their property; (2) people who are older, particularly elderly women, who may not have other access to seafood; and (3) his family. He also shares with friends, fellow hunters, and other neighbors. Another example of a typical sharing pattern was provided by a pair of shrimpers:

> This couple is in their fifties, and they go shrimping together for days at a time. From the shrimp they caught, they shared ice chests of shrimp with family members. They shared with her sister who lives in town, because when they need to evacuate, she lets them stay at her house. She also shared an ice chest of shrimp with her nephew who owns an equipment rental business. Anytime they need to rent some kind of equipment, he does not charge them anything, so it is a good trade as far as they are concerned.

The couple shares shrimp with their family, in part because it strengthens these bonds and creates reciprocal obligations. Here, these obligations cover both what might be considered normal family hospitality (staying with a sister during a hurricane) and what is typically seen as a contractual relationship (renting equipment from a business). The market value of lodging during a hurricane is considerable, and becoming more so.

As is clear in this discussion, sharing is not one thing. Sharing can be practiced to avoid waste, it can be done to cement a relationship, it can serve as a guise for what is really an exchange, it can be spontaneous (like Wendy's sharing of a fish), it can be routinized (like always bringing a cooler of shrimp to someone's house after a trawl), or it can be ritualized (like Tyler's Sunday gatherings or fish fries during Lent). Sharing in close networks can be almost involuntary or it may be rule based (for example, at some hunting clubs), and some foods are shared more than others. For someone with a satsuma tree that is producing heavily, sharing is almost a necessity to keep from either having rotting fruit around you or to committing to hours of squeezing and freezing juice over several days. In this context, not sharing would be seen as atypical or even strange. Unpacking the structures and layers of meaning in these sharing relationships will be a complex task for future studies.[6]

Exchange: Or How to Sell to Friends and Neighbors

Although some idealized views of subsistence exchange may reject the idea of any monetary involvement, we think that such an approach is not productive in the case of coastal Louisiana or in much of the contemporary world: it denies the complexity of the relationships and exchanges that we saw and consider to be within a subsistence framework. Rather than looking to money—its presence or absence—as the ultimate and only consideration for assessing whether or not a transaction is part of a subsistence practice, we find it more productive to consider whether a transaction's price point is primarily market driven or driven by social relationships or networks.[7] For example, you could say that Felicia's dad exchanged his wiring ability for access to the camp. He and the camp owner were friends, so he probably would have been invited to the camp anyway. But the wiring still functions as a contribution to the cost of the camp, a type of exchange. In some cases, like Arthur's, neighbors contribute labor, but they also pay cash for corn and tomatoes. Like sharing, exchange is not one thing.

Here are some examples of exchanges we documented that are clearly linked to subsistence practices:

> At Arthur's annual corn picking, friends and neighbors are called from a list compiled over years by word of mouth to come and help. Some do help, and most are sold bags of fresh corn at a below-market price. The money from

the sale of the corn is used to fund the rest of Arthur's gardening projects and cover the cost of the corn crop.

Many shrimpers reported having multiple prices: one for the dock, one for the shrimp factory, lower prices for the elderly, and a special price for certain neighbors and friends.

A crabber works a wage job part-time and gives his supervisor free crabs in exchange for flex time or time off to hunt or harvest at key times. That flex time is important for allowing him to keep his freezers full.

In each case there is either actual money or a money substitute represented here (time at work). And yet the key to understanding these interactions is not to focus on the money. The price in these cases is determined by social considerations. These below-market cash exchanges are about social networks, and the value in these exchanges is about more than the money. As illustrated in chapter 3, with the story of how David, the barber who traveled each week to fetch crabs and then resell them to friends and family in Thibodaux, defines the importance not by money because he's only earning enough to cover his gas. Instead, this cash exchange is about the relationships you build through those activities. There are layers of meaning: these transactions combine aspects of gift exchange and cultural values with the sales. A sale of corn is intermingled with a gift of labor. Crabs gifted to a supervisor help to maintain positive relationships at work. Sales and purchases of crabs are more about reputation and social networks than monetary gain. While sales may be needed to meet expenses (seed, gas, boat repairs, and other bills), they are also shaped by ethical and moral values that are socially and culturally grounded: these gift-sales are also about doing the right thing and being a good person.

Sometimes exchanges are more formal but don't involve direct cash. People barter. Many people told us about how their parents or grandparents had paid for ammunition or other essentials in furs or ducks, thus implying that this kind of exchange no longer took place. However, the practice of bartering is alive and well. For example, one shrimper told us that he paid for a child's haircuts in shrimp. One of our researchers, Mike Saunders, was engaged in multiple kinds of barter while working on the project. He was renovating a house in New Orleans in exchange for rent. In the course of repairs, he placed a broken air conditioner on the sidewalk. A truck drove by, and the driver asked if the air conditioner was for sale. The driver had just been fishing and offered to pay him in catfish (figure 23). A deal was struck.

Another participant told us about getting part of a house rewired in exchange for shrimp. Charles told a common story, a swap of docking space in exchange for seafood:

Figure 23. Researcher Mike Saunders holds up the catfish he swapped for a broken air conditioner. Photo courtesy of Mike Saunders.

I have a guy that parks his shrimp boat at my house. When he comes home [from shrimping] there's some things that the families over the years have said they like, and it's not necessarily shrimp. It might be squid. My sister-in-law ends up getting all the squid. He picks it up for her. Ice chests full of squid. Me, if I got to visit him at his boat when he comes he may give me a bucket of shrimp. Just enough to make a gumbo or a jambalaya, spaghetti, whatever it is that I want to cook with it, but it'll be a meal.

In effect, this shrimper pays for his docking space in shrimp and squid.

Roadside stands are a more cash-focused example of exchange. Driving bayou roads, you see dozens of signs of items for sale or actual stands. One fieldworker, Chris Adams, identified forty-four roadside stands selling twenty-seven products on a ten-mile stretch of two bayou side roads in Lafourche

Table 1. Items for Sale on Small Roadside Stands in Terrebonne and Lafourche Parishes

bell peppers	hay	potatoes
carrots	honey	pumpkins
choupic	mirlitons	rabbits
crabs, live / crab traps	mustard greens	satsumas
cushaw squash	oranges	scented candles
eggplants	orchids	sweet potatoes
field peas	okra	tomatoes
fig tree	onions	turkeys
garlic	peaches	winter squash
filé	persimmons	yard eggs

List of items for sale at small roadside stands during August, September, and October of 2013. This list does not include products sold at large, permanent stands, like the one in figure 25.

Explanations for some items in this list: Filé is a spice made from the dried and ground leaves of a sassafras tree. Commonly used in gumbo. Choupic (*Amia calva*) is often spelled choupique (but not on this sign) and is commonly called a bowfin or swamp trout in other parts of the US. As noted in the preface, satsumas are a type of mandarin orange, similar to a clementine. A cushaw squash, also called a cushaw pumpkin, silver-seed gourd, green-striped cushaw, or *juirdmon* (in Louisiana French), is a large crookneck squash (they can weigh up to twenty pounds) that has a long history in south Louisiana, particularly used to make cushaw pies (Elie 2009). "Field peas" can be a catch-all term for several varieties of legume. Commonly grown varieties include crowder, purple hull, and pinkeye. The only variety commonly available in stores is the black-eyed pea. These peas are grown in gardens throughout the US South.

Parish. Three other fieldworkers drove other roads and, in total, we documented thirty-seven items—not all food—for sale from people's yards (see table 1).[8]

This indicates a high saturation of food harvesting, production, and exchange taking place in small-scale sites. These are a visual reminder that subsistence practices and foods are widely dispersed in coastal communities, rather than being the singular purview of a small number of subsistence practitioners or specialists.

Stands come in varying sizes, and some larger ones can be associated with businesses or bring in foods from outside. The markers are often, in the words of one student, "little tiny signs" that a novice could miss completely. Some signs are small and weathered, the lettering barely visible, hanging from a mailbox or attached to a stake in a yard or by a driveway. Chris noted a sign for "Rabbits, spray painted onto plywood board, set upright on ground, but mostly hidden behind overgrown prickly pear cactus" on Highway 308. Another sign said "For sale, Rabbits [crossed out with duct tape], Fresh Eggs, Scented Candles"

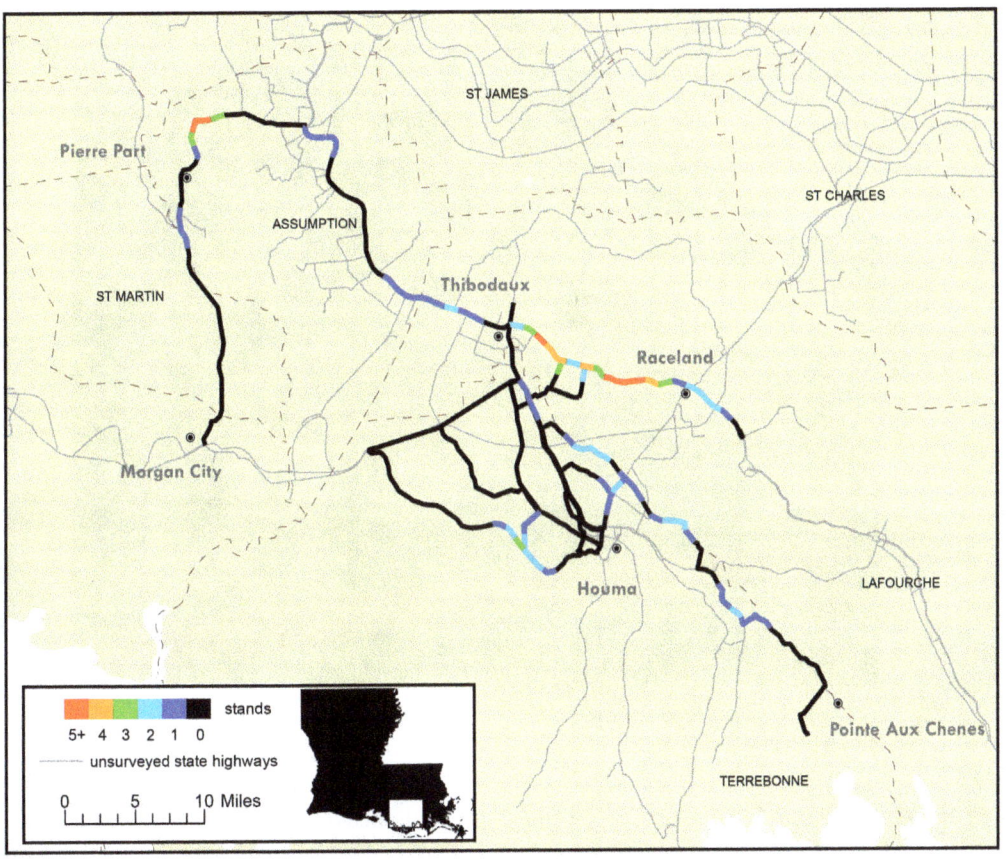

Figure 24. Map of roadside stands and roadside sale signs in August, September, and October of 2013, along major and secondary roads in mostly Lafourche and Terrebonne, going as far down the bayou as Pointe-aux-Chênes. Clearly, there are heavy concentrations along the well-traveled roads (Hwy 1 and State Road 308) that line Bayou Lafourche, from north of Thibodaux to the Lockport-Larose area, with as many as five stands in a mile. And again, there are almost continuous road stands from Schriever in Lafourche Parish, through the community of Bayou Blue, down to Bourg. Map created by Chris Adams.

with an arrow pointing down a side road and a phone number. Some signs are left up year-round, though the produce is only sold in-season. Another stand may have an empty wheelbarrow perched at the end of the driveway with blank signage, indicating a seasonal stand may be in operation at this location at some time of year, perhaps during citrus season. Hand-painted signs may accompany produce baskets and a simple money box or bucket, for purchases made on the honor code, as such small stands are often untended, though the owners usually live in a house nearby and check-in frequently.

Figure 25. Diamond Produce is an example of a larger roadside stand where most produce is grown by the sellers. Photo by Tiffany Duet.

Other stands are larger, more formal, with small sheds or buildings, as illustrated in the photo (figure 25). For example, a more established stand with a pavilion type roof and open-air walk-in stand enclosed with chicken wire fencing, announced "Diamond Produce. We only sell it if we grow it." In this way, it distinguished itself from commercial vegetable stands that resell food grown or harvested elsewhere. Another stand had a hand-painted sign for each item that was for sale (okra, cucumbers, gumbo filé, potatoes, onions, tomatoes, satsumas, bell peppers, sweet potatoes, garlic, pumpkins). Another fieldworker found larger produce stands, connected through family and social networks, that sell local produce.

The largest operations bring in harvested food from other areas. For example, several farm stands were owned by a larger area farm located in the Northshore/ Hammond area. Another produce stand was located in a permanent building with two vendors and contained items like commercial varieties of bananas. While most items sold in roadside stands were edible food items, researchers also recorded other farming-related sales, such as hay, or artisanal crafts, such as handmade candles.

Guns, Boats, Camps, and Freezers: Investments in Subsistence Production

Consumption like this entails costs. For example, costs include the time spent processing and preserving food and in learning the skills to do it properly. In Mike Saunders's field notes from a rabbit hunt, he describes how Richard Borne and the other hunters were watching as his grandson was struggling to learn to clean a rabbit properly (see chapter 5). Cleaning a rabbit is something you have to learn, just as you have to learn how to skin a deer, fillet a fish, devein shrimp, shuck corn, can tomatoes, make jelly, butcher a hog, and the hundreds of other skills necessary to preserve food (figure 26). Canned food must have appropriate seals to prevent spoiling. Food to be frozen must be properly stored so that it doesn't get freezer burn. Some foods can be thrown pretty much as is into the freezer while others will keep better if they are prepared. For instance, blueberries will store better if you prepare them carefully. First, you wash them and dry them completely (or let them air dry). Then spread them out on a cookie sheet, without the berries touching each other, and place that in the freezer to freeze. Once they are frozen, put them in a quart freezer bag and label them for easy storage. You need to use them within a year. Learning all of this takes time and practice.

Beyond time to learn, subsistence production requires investments in the necessary tools. These tools—ranging from fishing rods and reels, to shotguns, johnboats, and duck blinds (the list is endless)—are often acquired incrementally. People start young. In an interview, a Louisiana Wildlife and Fisheries agent said that most people who hunt and fish learn before age twenty. Nearly everyone we interviewed started as a child or teenager. Equipment is typically acquired gradually. A godfather (or *parrain*) purchases a first gun for a birthday gift. Then somebody hands down an old rod and reel when they get a new one. Buying some used crab traps may happen when someone has some extras they aren't using. A small tackle box gets started with just a few lures and then a few more. Christmas and birthdays are times to acquire a few more items—duck decoys, rifle scopes, or waders. Equipment gathered in this way is almost always worth more than its market value. Such possessions, like every gun Jerome has ever owned, store memories and recall relationships.

Parents, grandparents, or *parrains* (godparents) help also by providing access. In the example of John's camp (from chapter 6), his father bought it, along with a small group of friends, in 1964. His father and all but one of the friends passed away, leaving the camp in John's care. The lease to hunt was signed over to John, who now bears the costs for upkeep and maintenance on the camp, the hunting lease, boat maintenance, dogs, gasoline, ammunition, and rifles. He does not have the added difficulties of locating new property to

Figure 26. Learning to skin a deer: A group of men at Richard Borne's hunting camp showing Richard's son, Darren Borne—then a teenager—how to skin the deer he shot. Richard is in the plaid shirt in the back center. Photo courtesy of Richard Borne and Haley Metzger.

hunt on, buying into a new lease, or purchasing or building a new camp. Clearly this kind of material inheritance comes with its burdens, but it also means one isn't starting from scratch in figuring out how to access hunting and fishing grounds. People can also be creative at finding additional ways to make their hunting season productive. For example, John and his other camp members pay a portion of their yearly hunting lease fees by trawling for shrimp. They rigged a trawl net to one of their boats and use it during the spring and summer trawling seasons to earn extra money. In 2000, John wrote in his camp log: "The trawling brought a couple hundred dollars in the coffers. We've paid the lease and are now part of the landed aristocracy."

People also make significant investments of their own incomes. John, for example, says he cannot walk into a particular sporting goods store without "charging $100 on the credit card." In pondering the money he has spent on equipment, Serigny rejected the idea that his duck hunting could be measured by "return on investment." Even fishing, an activity that is less expensive for him, has involved a significant expenditure of money over time:

Fishing is exactly the same. I've got a bunch of rods on the wall in my garage right here. I have 5 or 6 rods, and each of those, you're going to spend anywhere from $20 to $40 on a good spinning reel and $40 on a good solid graphite spinning rod, so you know, that $80 times 6, that's $500 worth of fishing [gear]. And then I have a tackle box in the boat with at least $200 worth of fishing hooks, lures, floats, stoppers, bobbers, all the stuff you need to fish with.

Clearly, John can easily calculate the worth of the equipment on his wall and compare that to the value of the harvest. But he says you can't talk about whether or not such activities are "worth it" in those terms.

In the following interview, Carl and his friend Mitch, residents of north Louisiana who hunt up there and on the coast, discussed the waterfowl point system (which limits the number and type of ducks you can shoot in a day) and why some people might quit hunting, when duck hunting is not "worth it":

Mitch: About three years back, a lot of people quit duck hunting.
Carl: Well, they gave us . . . three ducks, wasn't it?
Mitch: And it actually went from not that long ago, in the late 70s, you could kill ten if you killed the right amount of ducks.
Carl: You were on a point system.
Mitch: But because of regulation, the duck count was down nationally, and it got all the way down to where you couldn't kill but three. And a lot of people quit duck hunting. Now it's back up to six and we're pretty content with six. It's worth the trouble and the expense. If you can kill six a piece, it's worth going through it.

Carl explained that for him, "going through it" means paying the $500 a year it costs to be part of a duck hunting club with eighteen members who collectively lease three hundred acres. Unlike John, who owns his own camp, Carl has a dollar-for-dollar calculation of the value of duck hunting.

Others have abandoned activities they calculate to be no longer "worth it." When asked, people most often cited duck or deer hunting as not being cost effective. For example, in the following excerpt Charles, a shrimper, explains why he stopped duck hunting:

My brother-in-laws go [duck] hunting. I don't. I used to go. I don't go anymore. They like to hunt. I don't. I used to, but it was too much trouble. You know the Wildlife and Fisheries put your quota so low that going hunting became expensive by the time you got all your gear and the gas for your boat and you launched it and you come back and you can only kill two ducks or three ducks or something. It costs you the same amount to go out and kill 15 or 20. So when they start limiting you down to just a couple ducks, it's kind of really expensive. It's not worthwhile.

This computation does not include the fact that hunters, like Carl, often have to pay to hunt on leased land. This is in comparison to fishing and crabbing which Charles decides are worth it. "We went crabbing because like $2 [for bait] and the cost to get there, you can catch a bushel of crabs, and the crabs cost $35/$40 [a bushel]." Here, cost per unit of food produced is taken directly into consideration.

Some explain that deer hunting will come closer to paying for itself than duck hunting. They point to their freezers full of deer meat as justification for the expenses of the season. Others, like Richard, note that people often choose equipment that makes hunting more expensive than is absolutely necessary. He points out that some hunters spend a small fortune on camouflage clothing, automatic insulating gloves, fancy rifle scopes, and more. A hunter like himself, however, just hits the woods in jeans, his old army jacket, and a rifle that he's had for more than twenty years. The stand on his property was built years ago and requires little maintenance. These days, his cost investments are almost always for his grandchildren: new rifles, small boats, and the membership to Club Brule.

And yet, even Richard has to decide each year if his hunting club is "worth it." To determine this, he said he calculates how much deer meat he needs to bring in to recover the cost of paying for his deer club membership. The deer displaces meat he would have normally purchased at a grocery store, saving him those dollars and allowing him to pay for his membership. Though this appears to be a straightforward—cost of investment versus value of product received—in actuality, he considers only part of the money spent. He noted that the membership fee does not cover the cost of the ammunition or the gas for the boats to get to the deer camp, all beyond the initial investment in guns, hunting gear, or boats. So his calculation has some sleight of hand, considering only one direct cost—the membership. When Shana pointed this out, Richard said, "Well, I'm going to hunt."

One thing everyone agreed on: hunting is more cost effective when people are able to put away and use more meat. For instance, as we saw in chapter 2, Jerome reported that he and his family rely almost entirely on wild harvests for their meat protein. They buy no meat at the grocery store, except for an occasional pork chop. And in their eating and cooking logs, the Dupré family also relied almost entirely on wild harvest provided by their son and grandson for their animal protein, including deer, wild hog, squirrel, and fish.

Figure 27. Jenny Bourg, a gardener in Terrebonne Parish, has learned to grow crops in raised beds because of the poor soil down the bayou. This photo shows a truck full of composted manure, which she buys in bulk in order to save money. Photo courtesy of Jenny Bourg.

Gardens also require investments. Particularly down the bayou, in places like Pointe-aux-Chênes, topsoil has been washed away, leached of nutrients, or contaminated by multiple floodings. Even farther up Bayou Terrebonne, Jenny Bourg, who wanted to garden like her grandparents, had to haul dozens of bags when she began with just two raised beds in 2017. As the garden grew, she and her husband built more raised beds, trellises for climbing plants, and paved areas for easy walking. She invests each season in soil additions like bone meal (to grow strong roots and provide food for the plants) and vermiculite (to help with both retaining water and drainage). In 2021, Jenny noted that she was buying composted manure thirty bags at a time in order to get a 10 percent discount (figure 27). As we noted in the opening of this chapter, Arthur

Bergeron's sales basically fund his garden. In order to garden at the scale he does, he uses the money from sales to pay for seed, field, fertilizer, as well as innovations, like building seed beds.

Production investments are not just about physical things. "You need more than a gun and bullets and a license to go deer hunting. You have to know what you're doing." Most people serve an apprenticeship of sorts by learning from older people. Their parents, godparents, grandparents, older friends, and older cousins took them out and showed them how to hunt. Jerome explained his own apprenticeship and how he is now teaching others:

> I used to go with a man that had the lease, he owned the NAPA store up the road. And he'd bring us, since I was a little boy, he'd bring us over there. We'd go hunting and all, but we was mostly their gophers. You know, "Go for this" "Go get that" "Go help him make that blind." But we were still able to hunt throughout the year with them. They'd still bring us. One day I was going out to the camp, out on the water that they own, and I had his grandsons with me, and I told them, "See how young you all are? And how old I am?" They said, "Yeah." I said, "That's how I used to be with your Papa." I said, I was the little boy then, and your Papa would bring us hunting. We'd go to the camp and help them build up blinds and whatever. And I said, "Now I'm taking y'all."

Here, there is an investment of time in the community's children. That time is spent teaching the children a valuable skill, but Jerome and others have told us that the time does more than that. He joked and said that he always "did hang around older people, you know. [. . .] And I guess that's why I'm not in all that kind of trouble today." He attributes this experience to why his life has turned out so well. Others told us the same thing. Part of the training is not only about the skill set of hunting, but about life training, absorbing (often nonverbally) an approach or attitude toward life. Recall how in chapter 1, Nicholls student Cory explained that his grandfather's shrimping lessons were what had shaped his brother into an admirable man:

> He remembers his grandpa always telling him how to live right. "He was teaching me how to be a man," Ryan said. He believes that is why he is the person he is today, a person with a traditional mindset.

In addition to life lessons learned during production—the harvesting itself—hunters, fishers, and gardeners create social relationships through these activities. People told us that their strongest and closest relationships, outside of their families, are with their hunting friends.

Another way to measure whether subsistence is "worth it" is to consider the value of having food that you provided. As we read in the student essay by Rory in chapter 1:

> The hard work and passion put into catching and cooking this food makes it taste all the better and everyone appreciates it more. It gives a whole different edge compared to just buying food from the grocery store.

For Rory and his family, part of the worth of that meal is knowing that his father and their friends, men who work full-time in the oil fields, dedicated their labor to bringing that food to the table. Of course, there will be other food, such as potato salads, vegetables, chips, and typically soft drinks and beer.

"Did I Say That Deer I Killed Saved Me Money?"

As we've shown throughout this chapter, subsistence practitioners constantly ask themselves whether or not their practices are "worth it." But hunters disagree about cost effectiveness or whether it should even be a consideration. John put it this way:

> If I had to compute the cost of a pound of duck meat, for me, I'd just quit. I mean, I can buy ribeye steaks way cheaper, even at $7.99 a pound, which I think is ludicrous. Even at $7.99 a pound I can buy ribeye steaks cheaper than I can get duck meat.

It's not that hunters don't think about cost, but rather that their calculations are layered, and rooted in the meaning and value of what they hunt. Here is a thoughtful reflection recorded by Jamie Digilormo with Carl as he muses on the role of deer and fish in his diet (and budget):

> **Carl:** See, I eat fish three times a month, maybe four. But I'm going to have deer meat—like I say, last night we cooked a big pot of spaghetti. Well, that's deer burger that we took out [to make the spaghetti sauce]. And I have deer sausage and deer burger, and we mix them together a little bit and make spaghetti or meatloaf, anything. There's nothing we cook that does not have wild game.
>
> **Jamie:** So, you think that this lifestyle of hunting and fishing has significantly impacted your grocery shopping and the amount of money you spend at the store?

Carl: Oh yeah! Tremendously! If I had to go buy meat—I mean, I love beef steak, don't get me wrong. I love a rack of ribs, but when you got to go and buy all that stuff? Your grocery bill will go up dramatically. Now, let's turn this around. Did I say that deer I killed saved me money? Probably not. I might have $2,000 in that deer stand that I killed him out of. I got another $1,000 in the rifle. I got another $1,000 in the deer leases every year. It's a no-no to try to put a price on it.

Carl makes a complex point. He both does and does not "save" money. He both does and does not see his hunting in terms of economics. True, he's not spending money at the grocery store, but he is spending money in other ways.

Scholars writing about the economic dimensions of subsistence in Canada have also wrestled with this ambiguity and complexity. In a 2016 essay on "theorizing the cause of the shadows," Colin Duncan returns to the relationship between our understanding of subsistence practices and the way we think about dominant economies. Dominant systems cast their shadows on other practices, making them hard to see clearly. He reflects, "the shadows are a complex place partly because they are cast by an unstable, complex mess" (Duncan 2016:366):

> When we try to think about subsistence, we need to be wary about opposing self-provisioning and its products to an overly simplified, allegedly monolithic commodity form. Of course, the commodity form essentially seeks to impose homogeneity upon heterogeneity. The commodity form, although itself a mere abstraction, tends to think like a capitalist—or seems to! *The outdoors, by contrast, is an extremely heterogeneous place, so subsistence activities are likewise heterogeneous and always will be.* (Duncan 2016:366–67, emphasis added)

The scholar and the hunter are both making sense of the relationship between the market and the harvest, what can be purchased (the commodity) and what is harvested though subsistence practices. Duncan, in other words, echoes Carl. The calculations are a "complex mess."

How much money you spend on harvesting and hunting depends, we have learned, on your family history and your connections, and the type of hunting or harvesting you do. "It's a no-no to put a price on it." Some people, like Carl, who have to drive four to five hours to the coast to duck hunt or fish for flounder, spend a great deal on transportation. In contrast, residents of Terrebonne and Lafourche Parishes might make very small personal investments, over time, to be able to maintain a subsistence lifestyle. Investments in subsistence might be part of their gifting system from the time they are children or teenagers. They might have older relatives who enculturate them over many years. People from those areas might have key members of their families that help them participate in subsistence activities even when they have to hold demanding forty-hour-a-week (or more) jobs and have to move up the bayou, farther from the coast, or even to cities like New Orleans, Baton Rouge, or Houston. Clearly, the costs vary widely. And the benefits can range from low-cost food sources to better health and nutrition, leisure time outdoors, family ties, personal identity, self-respect, and "bragging rights." Subsistence activities produce highly valued foods that cannot be purchased at the store. These foods fuel family gatherings centered around feasting on crabs and shrimp as well as finfish, duck, and produce from the garden. Perhaps more important is something that traditional economics does not account for—the value of caring for others and becoming a certain kind of person. They are learning to be people who contribute something meaningful to the world.

8

SELF-RELIANCE, CARE, AND MUTUAL AID

In the fall of 2020, the Neighborhood Story Project collaborated with the Land Memory Bank and Seed Exchange to host Botanica, a series of virtual gatherings centering Indigenous gardeners and healers to share their knowledge of the region's ethnobotany.[1] Monique Verdin, the director of the Land Memory Bank and Seed Exchange, is a multimedia artist, writer, and member of the United Houma Nation, who is working across the coast with people in diverse communities who are reclaiming their connection to gardening. Below are some excerpts from Monique's reflections on her own garden, mutual aid, seed exchanges, and a project to connect a network of Indigenous Women's Gardens.

So, in my garden right now, I'm growing goldenrod, which is just starting to turn yellow, and I'm very excited about that. I also have elderberry, which is still making berries and has flowers on it. So, elderberry's just such an incredible plant, not a native. It came from Europe, which was surprising to find out. I have this incredible maypop field. And the flowers, the leaves, the vine, the fruit can all be used. And it also grows like a rhizome. So it just starts coming up all over the place. It's been there a couple of years now. It's pretty great. And there also, I'm growing a bunch of greens right now and cucumbers, the mugwort, the white sage that Mr. Whitney gave me and then also some native fruit plants, mulberry and plum and persimmon.

But one really sweet, beautiful dream of a project that I've been really blessed to be one of the stewards of, is creating a network of Indigenous gardens with women led gardeners. And, you know, it's this decentralized kind of way of thinking. And *Bulbancha*, which is the traditional Choctaw name for New Orleans, meaning "place of many tongues" or many languages. This network of Indigenous women are coming from many different places. There is a core group of Houma women, but there's also, you know, Indigenous women that are coming from across the Southeast, thinking of these gardens that can also be able to support each other, and also the benefit of us being connected and being able to support each other. And also having seed banks or plant propagation places that—knowing we're facing these uncertain times—maybe my plant stock, my seeds, they disappear because I get wiped out by a storm. I know that there's a sister garden that's waiting for me.

And so, in this reclamation of our relationship with the land and our networks with each other, we know it's not good to keep all your seeds in one basket or one place. And how do we prepare and how are we strategic and how are we thinking about this on the front end? There's something there that feels very old and familiar and the mutual aid support that I think is going to be crucial—or is crucial, you know— for humans to really move forward.

You know, we always had somebody coming by the house when I was a kid bringing bundles of beans or filé or whatever. And I feel that [sense of community] in this old but very new kind of way, with this community of women that I've been able to connect with.

✦ ✦ ✦

This chapter tackles an idea that people stressed over and over again—hunting and harvesting is more than a way to make good food—it's how you make good people and good communities. Going further, that having access to those practices that help shape good people and strong communities is your birthright as a bayou person. Finally, that having this worldview brings resilience—the ability to survive and thrive in the face of great change. In her talk, Verdin called for a "reclamation of our relationship with the land and our networks with each other."

One of our first tasks in this project was to understand how coastal people themselves saw subsistence. As we noted at the beginning, the word *subsistence* itself was a stumbling block. To many people, it pointed to a practice that was, perhaps, backward or implied that all you did was "live off the land" and, thereby, did not participate in the modern world. Recall from chapter 5, Claudia Autin's concern that an interview with us would cast her family as characters out of a reality TV show like *Swamp People*. People were quick to explain that in order to truly "live off the land," a citizen of Terrebonne or Lafourche would have to flout existing laws and regulations. In other words, the catch limits, tags, permits, licensing requirements, and the existence of hunting and fishing seasons, limiting when you could hunt or shrimp, were perceived as making a fully subsistence lifestyle either impossible or illegal. As duck hunter John Serigny

"Subsistence"	"What we do here"
Poverty	Wealth or abundance. The necessary tools, equipment, and leases are often an expensive investment.
Food Needs / Lack of Choice / Scarcity	Homegrown and fresh-caught foods are highly desirable and linked to recreation, family, and feasting.
Backward / Primitive	Training and expertise; hunting and harvesting requires years of training to master and offers the chance to learn key values for a good life.
Lack of Wage Employment	Self-reliant and hard working.
Isolating / Living Apart from Society	Connected to community; sharing and mutual aid.

Table 2. Table comparing associations with the word "subsistence" to how people in the region talk about hunting, fishing, and gardening.

explained, feeding a family in generations past required far more harvesting than today's limits allow:

> My dad had eight brothers and sisters and then my grandmother and grandfather, so they had 10 mouths to feed. And my dad says he remembers washtubs full of innards from the ducks they would clean. The rest of the people in the community would share when they had a windfall like that. So that's how hunting started here. Kill as much as you can, when you can, because it might not be back tomorrow. Now, finally, we've gotten past that. Most of the young men who are hunting now were raised in an era where limits determined the number of ducks you could kill. But back then, when I started hunting in the '60s, if there were lots of dos gris,[2] you killed as many of them as possible. If there were lots of pintails, you killed that. You killed what you could when you could. It was the mentality of subsistence hunting.

Here John draws a bright line between how his parents lived and how he lives. They harvested primarily for food; his personal reasons are more complex. We reviewed dozens of discussions about subsistence, formal and informal, and compared how people talked about their ideas of subsistence versus what they see themselves as practicing (see table 2).

When trying to label their practices, people said, "what we do here," "our heritage," "living off the land," and "food you provide yourself." Many people explained that they might use several labels because people in coastal Louisiana did not necessarily see all hunters, fishers, and gardeners as doing just one thing. Sometimes you go fishing because you need to relax in the outdoors, sometimes because you want to hang out with a friend, and sometimes it's about needing to catch a lot of fish in order to supply a family gathering. And sometimes, as Monique pointed out in her talk, subsistence practices are about staking a claim to the land and shaping the communities you want.

Indigenous writer Enrique Salmón moves away from categories of "economics" or "recreation" to rethink such community-embedded practices as "eating the landscape," which he frames as coming to understanding a kinship with the ecology. "I realized a comfort and a sense of security that I was bound to everything around me in a reciprocal relationship" (2012:2):

> Eating a landscape is a socially reaffirming act. In the case of my family, whenever I partake of Eloisa's tamale recipe or my mother's way of preparing salsa, I am eating the memories and knowledge associated with those foods. (Salmón 2012:4)

As Salmón and others have pointed out, eating the landscape is also a political act, an assertion that access to healthy land and clean water, to the ability to provide (to some extent) for yourself and your family is essential to how we build healthy people and communities.

These ideas that harvesting and household production shape people are not at all unique to coastal communities. A recent study by Rauna Kuokkanen (2011) considers the case of Indigenous governance of subsistence communities and argues that household production and sharing manifest Indigenous worldviews:

> Indigenous economies such as household production and subsistence activities extend far beyond the economic sphere: they are at the heart of who people are culturally and socially. These economies, including the practices of sharing, manifest indigenous worldviews characterized by interdependence and reciprocity that extend to all living beings and to the land. (2011:217)

Clearly, this applies also to both Native and non-Native people in coastal Louisiana. Look at the list of traits people pointed to in defining "What We Do"—making choices that align with your culture and heritage, good food as a type of wealth, having skills and expertise to be able to produce and harvest food for your family and community—we can see the worldview of this community of practice. As we have seen in chapters on heritage, economics, family, feasting, community in clubs and camps, over and again people are telling us their values. We try to capture some of those values we heard is this list, in no particular order of importance:

> Respect your heritage and history.
> Share widely and often.
> See and appreciate the beauty in the woods, the bayous, and marshes.
> Eat good food, and place a premium on food someone harvests for you.
> Spend time with your family and friends.
> Consider yourself a caretaker of the practices you inherit and the place your ancestors called home.
> Learn to do things to be self-reliant—to hunt, to garden, to shoot, to make jug lines, to wire houses, to build a camp, to recognize and pick sassafras; to save and share seeds; to cook jambalaya.
> Enjoy your life.
> Teach what you know.

We now consider how people enculturate values through subsistence practices generally, and then look more closely at this worldview of what constitutes a good person and a good community, focusing on people who share, who practice self-reliance and environmental stewardship.

While we are exploring the centrality of hunting and harvesting to inculcating values, we want to back up and point out, as we have before, that we are not saying that all people up and down the bayou hunt, fish, or have gardens. This is worth emphasizing. Hunting and harvesting are also personal choices. For example, Glynn, the shrimper and sometimes alligator hunter (from chapter 1) who worked to keep subsistence practices central to his daily life, said none of his siblings or children were as interested:

> I have three brothers. I probably am the only one that went into it. The younger ones kind of tended to go out in the oilfield and stuff like that. I'm the oldest. I have three boys. One's a machinist. One runs a shipyard, but he does the alligator thing with me. And my oldest son, he works at a shipyard too.

When Audri asked Glynn if he had taken all of his sons out hunting alligators, he replied: "Yeah, I took them all out. Just like anything else, you know. Some will take to the ways and some won't." The idea is not that everybody has to hunt and harvest—particularly not as their main occupation, like Glynn does—but rather that hunting, harvesting, sharing, and all the values that come with those practices are part of the community. You don't have to hunt, fish, or grow a garden to be a good person, to care for, or keep your community, it just helps.

Hunting and Harvesting Shape Identity

Shrimping no longer provides a comfortable living for many people as it once did. But for many who continue in the trade, the work is part of their identity. In 2012 and 2013, researcher Audriana Hubbard was an anthropology graduate student at LSU conducting fieldwork for her master's thesis on the Blessing of the Fleet, a ceremony and community festival which happens at the opening of the brown shrimp season each spring (Hubbard 2013).[3] Many shrimpers told her how they felt that shrimping was a central part of who they are, both as individuals and as members of a larger (collective) way of life. One shrimper told her, "It's in your blood." Another shrimper, Louis, was thirty-six when Audri interviewed him, and he had been shrimping for more than twenty years. He is the third generation in his family to shrimp, and even as his wages decline and he seeks additional income sources, he still wants his sons to become shrimpers. From Audri's field notes:

> Louis believes that shrimpers "have an inner feeling" about being in the fisheries industry and that he's had that feeling all of his life and never questioned his career choice. "I always knew my destiny." Louis is a full-time shrimper, with two boats and a commercial permit. He shrimps almost constantly from the end of April through December, but he must earn enough to offset large costs, including $10,000 for commercial permits, thousands of dollars in gasoline for each long trip, and in 2012, the cost of a new boat, more than $300,000. Even when his shrimp hauls are good, which they usually are, the prices make for slim profit margins. So, from December to April, Louis takes as many small jobs as he can.
>
> Louis's love of shrimping is so strong that he wants to see it carry through to the next generation. He has a clear vision for what strong men do in his culture, including providing for a wife and children, looking out for the larger family, and self-provisioning. Louis sells his catch at seafood docks and retails directly to the public. He has customers who have been with him for ages, and even though his costs go up, he doesn't always raise the prices for his customers, even cutting deals with the elderly. He has a personal relationship with everyone he does business with, including the people who manage the seafood docks. Some years he makes up to $100,000 in two months, and other years he doesn't break even. Louis says the trick to remaining a shrimper is to be prepared for the lean times and pace out your spending as much as possible.

Shrimp are a constant part of his diet, almost every week of the year. And Louis gives away about 200–300 pounds of shrimp almost each time one of his boats comes back in shore. Just before our interview, he had donated 200 pounds of shrimp to the school band. The band used his shrimp and other donated shrimp and raised $35,000 selling $10-a-plate shrimp dinners. While he and Audri were chatting at the dock, a steady stream of people came up to say hello—friends who work on his boats, nephews and cousins, people from the community. Simply put, this is who Louis is—a shrimper who is working the waters that his great-grandfather trawled, providing for a family, selling to his friends and neighbors, and helping his community.

In this portrait of Louis, we see a person who completely identifies with his occupation, which to him is intricately linked to his sense of place and cultural heritage. And, importantly, he believes that this work shaped him into a good person for his family and community. Louis's perspective was common in our study, not only among shrimpers but among many hunters, fishers, and gardeners, who told us that their sense of place and the activities linked to place were ingrained in their identity. More than that, over and over people—from college students to grandparents—explained that hunting and harvesting were a key part of how you transmit values, build family and community, and generally create good humans.

When Jamie Digilormo spent a day crabbing at a public site on the western side of Louisiana's Gulf Coast, she interviewed dozens of fellow crabbers, ranging in age from twenty to eighty-five. People crabbing there identified themselves as white, Black, Cajun, African American, and Native American, men and women. She wrote this in her field notes:

> Hunting and fishing played a large role in their lives from an early age and thus helped to mold their self-identity. Being out in the natural landscape of southwest Louisiana brings back fond memories of spending time with family and friends, especially for those who have moved out of state. One elderly man in a wheelchair told me he had just driven down with some friends from Arkansas. He got a little emotional talking about his childhood days spent out there reading the marsh, studying the patterns of the tide and birds in order to be a successful fisherman and crabber.
>
> Most of the people I talked with grew up hunting and fishing with relatives and friends; they believe it to have had a major impact on making them moral and ethical people. For this reason, all had the desire to include their children in their hunting and fishing ventures. The men wanted to teach their sons, in particular, how to be men and take care of themselves in the world. As one father said about his teenage son, "He's a little too much into video games. I force him to fish and hunt as a taste of reality."

While many talked about hunting and fishing as a continuous thread, tying together different parts of their lives, others talked of returning to these activities after they were interrupted by other occupations. A charter fishing guide we interviewed had started working as a guide after many years working in other fields. Now he said he was finally "coming home." He told us he identified himself as someone who, from a young age, was shaped by subsistence activities:

> I guess I've been fishing 52 or 53 years. I remember as a kid my father, whenever we had time we'd go out fishing for perch or whatever. I remember one time my father, he had sent several of my brothers and sisters to a camp for eight weeks in North Carolina. And I was supposed to go with them. But because I was ill, I didn't go. What happened was I fell in the bayou and drank some bayou water; they think maybe that's where I got sick from. So he took me by myself and we went fishing. And we caught fish until we ran out of bait. We caught a lot. We even went into the middle of the lake, and we started picking out the big lily pads. He said he had always heard that little worms grew at the base of them. So we went out there and we got these white grubs at the base so we could fish some more. He won't remember that, my dad, but I do. As a kid, memories are made out of water. Fishing, it's a bonding time for parents.

After all these years, being out on the water still evokes a feeling of connection. He remembers the experience of fishing alone with his father—made all the more special because he grew up in such a large family. "It's a bonding time." It took him many years to find his way back to his childhood love of fishing. He considered becoming a charter fishing guide a way of "coming home." This story also turns on his experience of vulnerability (getting sick) and being cared for by a parent, the pivotal role of scarcity ("we caught fish until we ran out of bait"), and his father's resourcefulness locating and harvesting grubs under lily pads for use as bait—a practice stemming from traditional ecological knowledge ("he had always heard").

As this fishing guide's story shows, the centrality of subsistence activities to identity is not limited to people who make their living from them. Fishing was central to his identity first. It was later that he decided to earn a living from these activities. The guide explained the centrality of these practices as not arising from the job, but rather the job grew from his own values, which he saw as embedded in his identity as a member of the culture, primarily because he grew up hunting and fishing:

> It's something that's our culture here—we hunt, we fish, we shrimp, we trap, as a way of living. You grew up that way. You don't live here and decide to, at age 40, "Oh I think I'm going to start fishing." Most people grow up fishing.

Becoming a Person Who Shares and Cares

Sharing is highly valued. In her fieldwork with duck hunters, Annemarie found that almost everyone talked about sharing their bounty. From her field notes:

> Many duck hunters I spent time with reported that they shared their catch with older members of the community, those that previously hunted but were now physically unable to do so. One participant reported that he brought his catch to his grandparents' elderly neighbor because she has the traditional knowledge of plucking and preparing the bird. He, in turn, visits with her and they eat the catch together after she's prepared it. In addition to these informal sharing networks, the popularity of sport hunting and subsequently high numbers of catch, have led community organizations to develop more formal sharing networks. I met with South Louisiana representatives from Hunters for the Hungry, a national organization that develops annual sharing events for hunters who have no one to share their catch with.

If this sounds familiar, it's because we heard very similar feelings about sharing in chapter 7 from shrimpers Glynn and Charles, who said sharing was "no big deal," and from Wendy, who regularly provisioned her neighbors. Even more, the food logs showed sharing to be frequent—the rule, not the exception. In that section, we focused primarily on how sharing is part of a system of economics, while here we turn to the way people value sharing for its own sake and for how it makes them feel.

Mary Ann Griffin on Bayou Little Caillou grows a small garden, but, nevertheless, distributes much of the food to friends, family, and neighbors. She casually says: "We all give to each other." She illustrated by noting that her brother-in-law had recently brought her figs and she turned them into fig preserves. "I give some to all my neighbors." She feels this is an integral part of what it means to be in a shrimping community:

> It's part of our community, the sharing. When I have something, I make sure I have enough and I give it to all different ones. I cook and I give it to my neighbors and all. And then, like Jerome, if he has extra crabs he'll bring them to someone. All cooked and ready to eat. And all of my son-in-laws do that. They're all sharing.

Processing is often a key part of the sharing network. In the logs, rates of sharing for some families were as frequent as harvesting activities. Some people are key in networks for sharing and redistribution. If you look back at the food logs from the Dupré family, you can see Mrs. Dupré, like Mary Ann Griffin,

Figure 28. Helen talking with Patty in a front yard of peppers she planted for her family to eat. Patty's house, in the back left of the photo, is raised more than eight feet off the ground to prevent flood damage. Photo by Shana Walton.

as a person at the center of a hub, taking in raw harvests, and processing and redistributing them, often as favorite dishes or preserves.

Sharing is so highly valued that sometimes it is almost institutionalized. Many coastal Louisiana hunters participate in an annual "Clean Out Your Freezer Day" in late summer and donate their previous season's catches to local food banks. These catches include deer and waterfowl, including duck and coot. This event provides hunters the chance to share their catch with those in need, and to clear out freezer space in preparation for the teal, general duck, and deer seasons starting in the fall and heading into winter. Almost every hunter from this project emphasized the importance of making sure the animal does not go to waste, and many hunters, particularly older men, reported self-regulating below the legal limit to minimize their impact on the populations and reduce waste.

In other cases, both the gardening and the sharing are acts of affirmation, a way of reclaiming land that is washing away. Patty, who lived near the turnoff to go to Isle de Jean Charles, had to haul in dozens of bags of dirt because the soil is weak from years of being washed away by storms and salt-water intrusion. She turned her entire front yard into a huge pepper patch (figure 28).

Row after row of peppers. Patty doesn't even eat peppers—she raised them all for her children and grandchildren who love them.

We often found elders at the center of the work to share and redistribute. For example, Lora Ann introduced Helen and Shana to a group of elder women of the United Houma Nation who farmed a plot of land together in Pointe-aux-Chênes. They didn't think we needed to use their full names or take photos of them—"No, don't get me in that photograph!" Nevertheless, the two main caretakers, Mary Ann and Nattie, took the time to explain their garden to us. They grew a wide variety of vegetables, collecting limbs to build trellises, repurposing discarded buoys to create raised beds, building their own rabbit traps to keep the wildlife from eating up their hard work (figures 29–32). They also collected seeds from their crop each year and carefully saved them to plant the next year. In a plot about the size of a small house lot, those elder women raised enough to distribute food (some raw, some in prepared dishes) to a network of fifty friends and neighbors. All learned to garden from their families and were working to share their knowledge with younger people. The garden reaffirms their identity as people who understand the bayou and how to grow food, as well as strengthens their family, community, and tribal ties.

A growing body of work in medical anthropology is exploring the work of care, which combines practical, ethical, and affective (emotional) labor. The dominant health system focuses on individual health, hiding social dimensions of care that tend to collective well-being.[4] Building on work by Annemarie Mol and Maria Puig de la Bellacasa, scholars are noticing collective dimensions of care, including self-care, that is collective, rather than individualistic (Chudakova 2017). The work of the Houma elders caring for gardens and sharing their produce can be seen as a practice of collective care work. In the same way, others like Patty (who grew an entire garden to share for others) and Glynn (who supplies many neighbors with seafood)—help us to see how subsistence practices embody care as a collective action. We found that subsistence practices, in short, were routinely associated with the cultural values of generosity and attention to others, a way to ensure the well-being of the family and community. Sharing food is a form of care work that nurtures bodies and builds stronger communities.

Self-Reliance and Autonomy

Two other key values, self-determination and self-reliance, might seem at first glance, to be in opposition to notions of sharing and community building. And yet, people cited these over and again, saying it's important to "be my own boss," "provide food myself," "be able to fix things myself." People have a great deal of

Self-Reliance, Care, and Mutual Aid

Figure 29, 30, 31, and 32. Just-picked cucumbers lined up on the seat in Nattie's pirogue, which sits under a carport (figure 29). Mary Ann used the railings on her porch as a trellis to grow beans (figure 30). A flotation device left behind by workers cleaning up the BP oil spill was repurposed to grow tomatoes (figure 31). A homemade rabbit trap (figure 32). Photos by Shana Walton.

pride in their life skills. But it's important to recognize that subsistence practices build up personal autonomy through the capacity to give (and connect) to others. Self-reliance is not the same as self-sufficiency. While self-sufficiency reflects how state agencies have viewed the supposed "dependency" of lower-income people, self-reliance is a more holistic lens for thinking about people's capacity to take action and to make decisions about their lives in relation to others. We draw from Ashanté Reese's (2019) use of self-reliance to talk about community-based food systems in Washington, DC. Ashanté Reese theorizes "geographies of self-reliance" as an approach that "situates self-reliance in food consumption and production as a cultural, political and spatial framework for navigating inequality" (2019:11). Reese notes that self-reliance is manifest in everyday practices and in both individual and collective strategies. Her framing centers *agency*—community members' capacity to act and to make decisions about their lives—in the midst of "structural constraints of food inequities" (Reese 2019:8). Above all, self-reliance as a framework works to counteract ways of seeing communities as defined by what they don't have—by lack, by deficits, or by *nothingness* (Reese 2–19, 44–47; also see our discussion of the term *subsistence* versus *what we do*).

Our field notes are full of accounts of how area residents take pride in being self-reliant. In chapter 7, Wendy was proud that her son, only a teenager, was able to fish, duck hunt, and go crabbing to help fill the freezer. These life skills extend beyond subsistence. In chapter 5, we saw how Felicia and Joe used their own labor to build their camps, with help from friends. People told us about constructing chicken coops and beehives, wiring houses, and building pirogues. The elder women in Pointe-aux-Chênes built their own rabbit traps to save their garden. We also learned how Arthur has kept his grandfather's tractor going for forty years and even did the welding to build his own shrimp boat. At the back of Arthur's house were several tall piles of wood scraps, tin cans, pieces of metal (figures 33 and 34). These rough assemblages, which might look like trash piles to outsiders, are common around houses in the bayou region and serve as a site for recycling and reuse. In short, these piles are a not refuse, but a resource.

After interviewing a shrimper and crabber, Louis, a member of the Pointe-au-Chien tribe, who had recently taken a part-time job to earn a bit of money, Audri wrote, "Common among many of the shrimpers and crabbers I meet, he cites working his own hours and being his own boss as the better aspects of the shrimping business." He didn't much care for the part-time job.

In her field notes, Audri heard Louis had put it even more strongly:

In the past, Louis gave up shrimping, and using his boat pilot license and some extra training, Louis got a job as a captain of a tugboat. He "made good money" and got to

Figure 33 and 34. Arthur Bergeron's recycling piles. A pile of No. 10 cans from a local school cafeteria that Arthur uses to place around baby tomato plants (figure 33). Two piles of brick (one under some brush) left over from past projects (figure 34). Photos by Shana Walton.

travel the region. But, "it was not my happiest time." He explained, "It was almost like a prison term." He had an aversion to having a fixed schedule in which he was regularly reporting to other people. But he still needs additional income, so he does carpentry, plumbing, and electrical work in the off months.

Two things we will note from this: First, Louis saw being a tugboat captain as "almost like a prison term," even though he was still working outside and on the water. Second, Louis had enough skills to earn money as a handyman. He knew carpentry, plumbing, and wiring. Most people we talked to hold more traditional wage jobs. For them, building your own camp and providing your own food affirms, in some way, that you can provide for yourself. And, as Rory said in chapter 1, the pride in hunting or growing the food yourself just makes the dishes taste better.

At times, connected to these values of self-reliance and self-determination was a substrate of ideas about masculinity. Here is another passage from Audri's field notes with Louis:

> He has three children, one daughter and two sons. At the time of the interview, the oldest son had a disability that would make it difficult for him to work on a shrimp boat. His younger son was a teenager and Louis looked to him to take up the vocation. He was concerned with transmitting his views on gender identity as well. "I'm a manly man," he reflected. He and his wife disagreed about whether their teenaged son should be allowed to play baseball or plan for college. Louis did not want his son spending a summer on a baseball diamond, but instead out on the boats: "I'd like to see him be a man."

For Louis, a ball team could not teach his son the lessons he needed to grow up to be a strong man. Remember Cory's essay in which he says his brother Ryan learned "to be a man" through the shrimping lessons of his grandfather.

Harvesting as masculinity is not a new idea (Ownby 1990). Masculinity intersects with racial and ethnic identity in complex ways. We certainly heard references linking being "Cajun" with masculinity. For example, in Annemarie's interviews with alligator hunters, a man in his sixties is called a "true Cajun, living off the land" and another hunter in his forties is referred to as a Cajun who was "the real deal" by other community members. This is significant, because in both cases, community members shifted their labeling from affinities, parentage, or community to abilities or practices, a way of linking the label to a form of respected masculinity. In another example of this linkage to "Cajun" (but using the sometimes-derogatory identity label "coonass") Richard shared a story about his son's experience in the marines. When his son participated in survival training and was left on an island with other recruits, he wound up teaching them how to trap, fish, and build a shelter. When asked how he knew so much about living off the land, the son replied: "I'm a coonass from Louisiana."

However, plenty of women participate in hunting and harvesting, as we have amply documented. Recall that Joe said his daughter got her first gun before she started school, a "20 gauge crack barrel" given to her by her grandfather. One of the first hunting photos Joe showed us was a picture of his daughter, age twelve, standing beside the first deer she shot. John talked at length about how he and his wife had duck hunted together. In our interviews with the three Houma elder gardeners, they all were avid fishers, and some of them had spent their fair share of time in a cold duck blind.

Environmental Stewardship

A common sentiment among coastal subsistence practitioners is that they are the true stewards of the environment. But that does not mean they feel their interests align with environmental movements in general. The larger environmental movement has historically often rhetorically advocated for conservation of natural resources, landscapes, and nonhuman species. In practice, however, some of the resulting policies—such as the creation of conservation areas, national parks, and preserves—have displaced the humans from places where they historically hunted, fished, gardened, or harvested plants (Kuokkanen 2011:223). Other campaigns urge consumers to preserve fisheries by avoiding purchasing specific species. Elspeth Probyn points out the irony of the "Save the Fish" approach to marine sustainability, because "of course in doing so we say no to millions of fishers" (Probyn 2016:31):

> In the public realm, there is little concern for commercial fishers. . . . Fishing for a living is portrayed as pillaging the sea for immense profits, whereas recreational fishing is seen as a benign activity despite its practitioners catching millions of tons of species, including valuable ones such as lobster and abalone. (Probyn 2016:39)

Again, we see the very categories used by conservation policy (commercial, recreational) marginalize or erase subsistence fishers who are more-than-recreational and who don't fit the commercial slot either. These fishers are often not even considered. In caring for the more-than-human, Probyn urges, we would do well to include "fishers, regulators, and indeed even the consumers of fish" (2016:31, 33).

Conservation issues and environmental knowledge were frequent topics in our interviews. Some people, like John, are actually longtime members of conversation groups. In his case, he has participated in Ducks Unlimited, a group dedicated to preserving waterfowl habitat—wetlands and waterways—with a focus on providing game hunting. John has been involved since the 1980s, not long after the first chapter was founded in Lafourche Parish. He is proud of the money his organization has spent for wetland restoration and the education they provide on safe hunting skills. John was one of many people who bemoaned the overharvesting witnessed in their childhoods.

Other people thought of their clubs as a type of conservation organization. Conservation was a frequent office conversation topic between Shana and Richard Borne (see the chapter "Camps, Clubs, and Leases"). While Richard thought the national environmental movement in general was about politics, not a genuine effort to restore land or help animals, he was himself greatly concerned about wetland loss and conservation in general. Richard pointed out that his hunting club set their own limits on does, a lower limit than the state, in their role as good stewards. Richard also said that one of the reasons he wants his grandchildren to learn to hunt is to "teach them conservation." Learning how to hunt is, for him, inextricably connected to learning how to take care of the environment. Here is how Mike Saunders documented it in his field notes:

> Richard also talks about marsh deer being pushed up into the swamp. This is important not only for showing how hunting is deeply connected to environmental knowledge, but also for showing how—in part—the meaning of subsistence practice today in coastal Louisiana is tied up with this knowledge of wetland loss and land loss and saltwater intrusion.

Richard's sentiments are widespread among the hunters, fishers, and gardeners we met. People were in agreement that, while there are some harvesters who are ignorant of wildlife problems and land loss issues—particularly, they point to "reckless youths" or people from out of state who they said have no investment in the region—the majority are passionately concerned about the environment, animal welfare, and sustainability. Our study participants told us that in order to have a right to make rules or even statements about the environment, a person should have deep knowledge, such as comes with living in the region and learning to hunt or fish. In other words, they believed hunting and harvesting had, in fact, made them true guardians of the environment.

9

CONCLUSION

Annemarie and Shana were at a booth set up during the 2012 Chauvin Folk Festival, held on the banks of Bayou Little Caillou, with handmade signs that said "Tell Us Your Story." A man came up and said he could easily tell us about the importance of hunting and harvesting: "We fish in the summer; we hunt in the winter; and sometimes we go to Houma to work."

✦ ✦ ✦

Louisiana subsistence practices had never been studied systematically prior to 2011 when we were asked by the Environmental Studies office of the Bureau of Ocean Energy Management, Gulf of Mexico Region, to develop a preliminary study of these practices. Early into the project, Helen met James Fall, the head of research in the Subsistence Division at the Alaska Department of Fish and Game, during a meeting of the Society for Applied Anthropology. Helen was apologetic about the experimental methods and preliminary nature of our findings, being mostly story based. Not to worry, he reassured her. "When we first started studying subsistence in Alaska, the only numbers in our reports were the page numbers!" He laughed. But he was saying something serious: that it's important to do the ethnography and understand the meaning of subsistence practices before trying to measure and quantify them.

With that in mind, we would like to return where we started, with a reevaluation of how to envision the concept of subsistence and subsistence practices. We proposed that what we see in coastal Louisiana leads us to think of such practices as emergent (that is, visible in the moment depending on the participants and actions), as multichanneled or hybrid (as opposed to simple or pure), as having goals other than profit (but not necessarily independent of revenue-generating activities), as contributing to food, pleasure, and personal identity, as linked to community, family, and heritage. Our ethnographic account includes snapshots of how people use subsistence as food, as part of their economic system, and to simultaneously make themselves into people who are connected (i.e., through sharing, family, community, heritage, and place) and autonomous (i.e., independence, self-provisioning, and self-reliance). Ultimately, we argue, for many people in our region, subsistence practices are both what we do and how we talk about what we do: how we take care of others through sharing food and how we raise children and make (good) people and communities through producing and sharing food. As we end our story, we want to make connections to two larger conversations that we see as connected to our subsistence work: (1) food sovereignty and (2) happiness and well-being studies.

From Food Security to Food Justice and Food Sovereignty

Early on in our project, we tended to think of the significance of subsistence practices and subsistence foods in terms of food security. After all, we were asked to work on the project in the aftermath of the 2010 oil disaster when many had experienced the loss of their subsistence foods (e.g., fish, shrimp, and crab). Because those losses were not documented, affected people had been unable to receive any compensation for those losses that were intensely personal and social as well as financial hits. Scholars, activists, and policy makers who work on food security are especially concerned with people having enough to eat and having access to fresh, nutritious foods. We considered whether down-the-bayou communities could be seen as food deserts (they are). As we worked on our report, we increasingly began to think about food justice, that is, the structurally unequal access to foods faced by many people. And we are increasingly thinking of this topic in terms of food sovereignty. Those working on food sovereignty are concerned with self-determination, advocating for developing (or preserving) ways for people to take charge of their own livelihoods.

For those involved in subsistence practices, not only hunting and gathering, but also sharing are tied into the human capacity to take care of ourselves and each other as members of social worlds. As part of a food-sharing network, families are practicing self-reliant foodways. We have seen through these chapters that subsistence practices create both a sense of independence and a sense of collective self-determination. Nancy Pottery, for example, writing about the Nipissing Nation's fishing practices in Ontario, Canada, explains that for Indigenous communities like the Nipissing, managing their fishery is a crucial part of enacting their sovereignty as a people (Pottery 2016). Indigenous fishing practices, like those of the Nipissing Nation, were often misunderstood by settlers and colonial officials and later by Canadian policy makers and administrators. The ways in which "subsistence" fishing was defined—as an economic practice in contrast to commercial and sports fishing—distorted the ways in which aboriginal people engaged in fishing for food and overlooked their historic involvement in long-distance trade. "Subsistence," "commercial," and "recreational" were not categories of fishing that made sense in Anishnaabe language or worldview (Pottery 2016:149).[1]

In fact, William Knight has written that fisheries categories of subsistence versus commercial are "largely administrative fictions designed to describe and regulate our interactions with nature, but fictive formations *with immense power* to construct a social order or moral economy within fisheries" (Knight 2009, cited in Pottery 2016:157, emphasis added). In other words, the administrators saw the fisheries in a way that made sense to them, and enacted rules that enforced that reality. The implications of subsistence versus commercial versus recreational in a Louisiana context need to be explored further. Many avid fishers on the coast talked to us about how redfish, once a subsistence food, was reclassified a game/sports fish after the Cajun food craze went national, as popularized by Louisiana celebrity chef Paul Prudhomme's signature dish Blackened Redfish in the 1980s. Currently, Louisiana regulations limit recreational redfish catches to "five redfish per person per day and each fish must measure more than 16 inches" (Masson 2017). The redfish population has rebounded as a result of catch and size limits, and the fish is now available to recreational fishers. But the limits mean that the fish can't be used to fill anyone's freezer. Subsistence practices were not taken into account. Categories like commercial and recreational fishing, and the setting of priority uses, limits, and licensing are critical issues in determining who gets to experience (or create) meaningful food sovereignty.

Well-Being and the Anthropology of Happiness

A central part of our book focuses on how, through gardening, fishing, and hunting, people come to feel good about themselves and make meaningful connections with others. To consider this further, we turn to the concept of *well-being* that has come into the social sciences from economics. The research on well-being emerges from economists' realization that metrics like income or wealth are unable to measure "what makes life good" (Mathews and Izquierdo 2009:13). Well-being can be seen as a combination of objective and subjective dimensions—measures both of how an individual perceives happiness and social and cultural indexes of life traits (see also Thin 2012). Anthropologists who study happiness and well-being are especially concerned with social and cultural dimensions. Walker and Kavedzija (2015) urge anthropologists to consider the links between happiness and the good, which we might think of as the relationship between *feeling good* and *doing good* (referencing the Greek concepts of *hedonia* and *eudaimonia*). Their research suggests that happiness is linked to other key human dimensions like relatedness (or kinship) and moral obligations to others (or ethics). They suggest that happiness is both "highly relational" and "other-oriented" (Walker and Kavedzija 2015:15). That is, well-being and happiness cut across the dominant ideology of individualism in the US, which places pressure on all of us to "get ahead," maximizing our individual earnings, pleasure, or advancement. "Happiness seems strikingly well-suited as a starting point for inquiring into what gives lives a sense of purpose or direction, or how people search for the best way to live" (Walker and Kavedzija 2015:16–17).

Other scholars have explored well-being in relation to environmental conservation. Environmental concerns, for example, have emerged as key to well-being among some Indigenous communities in the Andes-Amazon regions of South America. Collaborative dialogs between the communities and the Field Museum of Natural History in Chicago reveal that "well-being is measured in a more integrated way that places economic satisfaction within a broader context of social experience and their natural surroundings" (Wali 2012; Wali et al. 2017:6). To understand well-being in Indigenous communities, Wali and colleagues consider "the balance between humans, other life forms, and supernatural beings, and a moral dimension that regulates relationships, especially across generations" (Wali et al. 2017:6).

A wealth of studies has documented specific links between fisheries and well-being. One important collection was published in *Human Organization* by Patricia Clay and Courtland Smith (2010). As Shirley Fiske wrote:

> But most fisheries researchers have found that there is great satisfaction and adventure in fishing (commercial fishing) and fishermen feel more happy with their jobs and life than if they were doing something else. (Fiske, personal communication, June 2, 2022)

In Puerto Rico in particular fishers don't talk about what they do as "subsistence" and yet they assemble livelihoods in ways that resemble residents of coastal Louisiana, with wage work coexisting with commercial fishing, self-provisioning, and sharing (García-Quijano et al. 2015; Griffith and Valdéz Pizzini 2002). In their study, Fiske and Callaway (2020) summarize this body of work noting that these studies show how on any given day a fisher may be engaged in subsistence, recreation, or market fishing. In addition, they note that these studies show how "occupational multiplicity" is common with fishers also working in local construction, small-scale agriculture, and providing fish to tourist restaurants:

> [Garcia-Quijano and colleagues (2015)] make the argument that juggling these sources of production can bring changes to families' sense of well-being and identity, belonging and self-worth (Griffith and Valdes-Pizzini 2002). They show that part subsistence, part labor, part selling for cash is the norm, combining the sale of seafood with subsistence fishing for one's family consumption, and that these activities are a source of self-worth and identity for the family. (Fiske and Callaway 2020:180)

The scholarship from Puerto Rico resonates with our findings. Louisiana's subsistence practices are distinctive, and yet their links to well-being resemble those found in other coastal communities with an abundant seafood harvest.

Recent work by geographers on "community economy" suggests that growing, harvesting, and sharing fresh, delicious food is central to how people build relations of "mutual care, interdependence, recognition, collective wellbeing, and a sense of place" (Community Economies 2019:56). Advocating for the importance of research on these practices, they write, not only involves documenting and creating inventories of what people actually do, but raising awareness of these practices, bringing about "shifts in economic subjectivity" (Community Economies 2019:60). In other words, documenting what people actually do—real economic practice—can bring about a shift in awareness around what they value. Specifically, how we value specific resources, exchange, and relationships, how we become connected to each other, to place, and to other living beings. This is what abundance and well-being look like.

Here is where the concepts of well-being and happiness seem especially helpful for considering the social, economic, and political importance of subsistence practices in coastal Louisiana. Crab boils, deer hunts, camp weekends, sharing food with neighbors, building duck blinds, road fishing, constructing a boat camp, fish fries—these practices are not just about the food. Instead, people talked about the intersection between feeling good and doing good.

When we first considered Mrs. Dupré's pleasure in making oyster spaghetti, we discussed how much of what makes her foods meaningful comes from

Figure 35. This poster, part of a promotional campaign for tourism in Lafourche Parish, captures the idea that subsistence activities create family bonds and help raise children with good values. Created by DigIn! Lafourche, which can be found on Facebook and other social media outlets. Image in the public domain.

knowing the relationships that are built into making delicious food for herself and for her family, so that in receiving oysters, serving boiled shrimp from her son, sharing satsumas from her trees, and buying garlic bread from her grandchildren's school—what connects it all is not the source of the food, though this also matters, but how it connects her to people she cares about.

When shrimper Glynn Trahan talked with us about sharing part of his catch with others, particularly the elderly, he made it clear that part of the value of shrimping and crabbing was this ability to take care of others: "Like just now I was talking, when you called me, a lady was telling me how much she likes soft-shelled crab. Well, if I go shrimping and I catch soft-shelled crab, I'm going to bring that lady soft-shelled crab. I know she desires it. She don't get much of it, and she would like to have some."

What goes into creating these sharing practices is much more than the skill, the labor, and the abundance of natural resources. Participation requires tremendous environmental knowledge. It also stems from being able to organize your life so that subsistence is possible and within reach. To do this, one needs to be part of a larger community of people who have organized their lives in the same way. Members of this community share values around the appreciation of self-provisioning and sharing delicious fresh foods. Together, these elements create a sense of self-reliance, autonomy, and self-determination. These practices further a sense of being both a good person, one who is strongly connected to other people, and who is able to act on the world and in their own lives. Collectively, members of subsistence communities practice food sovereignty as they work to take care of themselves and each other. Families who have gathered around a crab boil or a fish fry take pleasure in not only the taste but the knowledge that every bite draws from an assemblage of practices—a way of life centered on food that not only tastes good but anchors the good life. As we quoted Jacob at the beginning of chapter 5, "crabs = happiness."

The feelings of connection that we have emphasized in our research are the connections bayou residents sustain with each other and with place—a connection to the land, to particular bodies of water, to particular foods, and to living beings. Bayou communities that generate and sustain subsistence practices are multispecies communities: they include alligators, duck, shrimp, crab, fish, and other life forms—pecan and satsuma trees, the palmetto harvested for making baskets, sassafras trees harvested for filé and the insects and other pollinators. We look forward to other future studies that will center those relations among living beings and the land.

Subsistence Futures

We conclude this book with the awareness that this study has captured only one place at one moment in time—a slice of subsistence practices in one region and during a period of three years (2011–2014). Even as we write, the landscape of subsistence practice is changing rapidly, and it is not being overly dramatic to say that subsistence is under threat. In south Louisiana, coastal land loss is an immediate threat to the lifeways and communities described in this project. For some, like members of Indigenous communities, identity itself is at risk, as subsistence practices form a central part of their experience as Indigenous or Native people.

We were overwhelmed by the response to this project. More people wanted to participate than we could possibly accommodate. This desire was spurred by interest in the topic, but also by a sense of urgency on the part of people who practice subsistence. Many were keenly aware of the rapid land loss, saltwater intrusion, and loss of habitat for plants and animals. They see these processes accelerated by hurricanes and human action which pose a threat to coastal harvests. Others were acutely aware of the possible decline of subsistence activities and the need to introduce young people to fishing and hunting so that these practices could continue.

We wanted to share the tone from one of our early community meetings. Before we even had a team in place, the volunteer firefighters from Bayou Little Caillou agreed to join us in 2011 for one of our first focus groups, both to discuss their own practices and to help shape our hunting and eating logs. A member of a deer camp was worried about land loss being so great that there would not be sufficient habitat for deer. People talked about shrimp prices, the global shrimp market, and cheap shrimp from overseas. Older firefighters remembered their parents trapping in the winters and had lived long enough to see the disappearance of this practice. Another participant, passionate about fishing and preferring to eat fresh-caught fish, said she felt a deep sense of loss when, in the aftermath of the 2010 *Deepwater Horizon* disaster, she had to stop eating Gulf seafood. Because she preferred fresh-caught fish, she never turned to the grocery store. She just stopped eating fish.

Without a doubt the most intense worry was that a confluence of changes and pressures—including shifts in lifestyle, habitat loss, increased costs, loss of access to existing land, people relocating—would mean their children would not experience the lifestyle they grew up with. At the fire station, many saw a bleak future, and one man said:

> I grew up in it, and I'm afraid that there's not as many people doing it now as there ought to be. I mean, you got to have your kids growing up hunting and fishing, or it's going to stop.

Another man in his twenties said he had a one-year-old child. He talked about how much he loved his young son and wanted this child to have the rich experiences he had known growing up. He put his head in his hands and began to sob. The grief, he said, is his fear that the world is changing so quickly that he will never fish and hunt with this child. His sadness, thinking that he was the last bayou generation to hunt and fish, was devastating.

People living in coastal communities face multiple forms of precarity. Precarity—a concept that folds in notions of vulnerability and insecurity—is defined by anthropologist Anna Tsing as "life without the promise of stability" (Tsing 2015:2). It's a word that speaks to living with multiple forms of uncertainty, including economic, political, and environmental change. In Louisiana, coastal residents face the ups and downs of the oil and gas industry, they confront environmental crises like oil disasters, with long-term impacts on food sources, health, and livelihoods, and the increasingly unpredictable storm seasons—with hurricanes that form and strengthen so quickly, that it is difficult to prepare. As Monique

Verdin pointed out (chapter 8), subsistence practices—harvesting, feasting, sharing, and checking in with family, friends, and neighbors regularly—form an organic pattern of mutual aid that helps bayou communities face challenges of economic and environmental uncertainty. Some might say subsistence makes coastal communities more resilient (Browne 2016).[2] Others, area residents, say their skills and experiences make them more *tenacious*.[3] If the exchange networks that grow up around food sharing also play a role in storm preparation, evacuation, and recovery—and we believe they do, though our study did not focus on this—then subsistence practices would contribute directly to help people live with hazards of hurricanes and floods. They might also make someone reluctant to move away, both from the resources of bayou, marsh, and ocean, and from the social networks. These issues are central to conversations about community relocation and climate migration. You can relocate a house, but how do you relocate entire social networks? And how does moving farther from the marsh and bayou affect your ability to participate in harvesting and sharing activities that are central to your identity and your well-being?

For some coastal Louisiana residents, phrases like *climate change* and *global warming* may be closely tied into partisan politics. But topics like hurricanes, flooding, and land loss are not. Everyone shares the increased risk, and the experience of precariousness of life in down-the-bayou communities, and increasingly in the up-the-bayou towns like Houma that have long served as destinations and as places of refuge during a storm (Parfait 2019). Berkeley sociologist Arlie Russell Hochschild (2016) has written of how rural and industrial communities in southwest Louisiana, living in the shadow of oil and gas industries, sometimes adopt political stances against their own interests.[4] In Terrebonne and Lafourche Parishes, several hours to the east, we found a more complicated picture. Particularly at the state and local level, people had a complex understanding of the urgency and the governmental actions that might be necessary. Residents are remarkably united by their concern for the environment, even when they have different visions of the future.[5]

Much of the environmental knowledge that is held by residents of coastal Louisiana who are deeply involved in subsistence harvesting and sharing—their perspectives on water, land, and the creatures that live there—is unsentimental and unromantic. For this reason, their experiences may be overlooked and underestimated by environmental activists who may be tempted to project their moral and philosophical orientations onto local landscapes. The Louisiana license plate proclaiming the state to be a "Sportsmen's Paradise" obscures as much as it reveals about the ways in which hunting, fishing, gardening, and sharing practices are a part of everyday life for those who fish for food as well as pleasure.

In our book, we provide a sketch of subsistence practice and a foundation for residents and scholars to build on and debate. We emphasize that for many people of all ethnicities who live down the bayou (or who grew up there), subsistence practices like fishing, shrimping, hunting, and gardening are an important part of their family heritage and continue to be part of their experience of place. From the opening of shrimp season to summer crab boils and winter gumbos, subsistence foods provide an occasion (and a good draw) for family gatherings, fill the holiday table, and provide tasty, nutritious food for many everyday meals. For many coastal Louisiana residents, knowing where your food comes from, and who caught, hunted, or harvested it, is a big part of what makes a mess of shrimp or a duck and sausage gumbo delicious and meaningful.

Although most Americans do not hunt and fish regularly, here in coastal Louisiana we are struck by the ordinariness, pervasiveness, and density of hunting and harvesting practices in the everyday lives of coastal residents. Hunting and harvesting activities are, simply put, important to people. What specifically do we mean by "important?" What we have seen is that people set their annual calendars around dates for hunting and harvesting. People teach these activities to their children and grandchildren. People make monetary investments—sometimes quite large—in equipment that allows them to hunt, fish, or garden. Items related to hunting and harvesting—perhaps a fishing pole or a nicely wrapped box of hand-shelled pecans—are treasured gifts. These are the activities people use to anchor many social events, from family reunions and weddings to weekly crab boils. People invest these activities with meanings, asserting that correct participation tells you who a person is and shows their values. Self-reliance, and being able to take care of yourself and others, is central to well-being and happiness. Narratives about hunting and harvesting are common and create community. In short, we found that people in southern Louisiana find subsistence activities to be tied to personal, family, community, social, cultural, economic, and aesthetic values. Important and pervasive, subsistence practices are like water in the Mississippi basin, both extensive and particular in any given moment.

As Harry Luton, the BOEM official who initiated this project, said so elegantly during his remarks at the 2019 Society for Applied Anthropology meeting, "Subsistence is not a thing. It's an existential condition. It's an expression of our humanity."

POSTSCRIPT

HURRICANE IDA

Lora Ann Chaisson is selling her house in Pointe-aux-Chênes. She was raised living down the bayou; her relatives all live in down-the-bayou communities. But she has made the hard decision to leave. The storms have won. Hurricanes Isidore and Lili (2002), Cindy, Katrina, Rita (2005), Gustav and Ike (2008), Barry (2019), Zeta (2020), and, in August 2021, Ida, which had 150-mph winds and was one of the strongest hurricanes to ever make landfall in the United States. Her house, raised eleven feet in the air to protect it from storms, suffered serious damage in Ida, and, now, to complicate matters, her foot requires surgery. If she stays, for months she would have to find a way to walk up the sixteen steps to her front porch. She writes:

Right now, I've made the hard decision to leave the bayou. It's not an easy decision. When I signed the sales contract, I thought, What am I doing? My family is from here. I wish I could keep my house and the land. But I can't. I can't even find anybody to come down and repair the house. So I have to sell. I love fishing; I don't know when I'll be able to fish.

I will have been in this house 21 years this spring, and I've lived through 10 hurricanes that hit here or nearby. And this is the eighth time I've repaired my house because of a hurricane. And now this time, after Ida, it's so depressing. It's just so depressing. To see everything around you is just gone. The house next door is gone. The house across the street is gone. I mean, totally gone. Another house a bit further down is heavily damaged. The Island [Isle de Jean Charles] is flat now. All up and down the bayou is flat. I've never seen this amount of damage. I worked helping with disaster applications after Katrina in Plaquemines Parish and St. Bernard Parish, and I saw a lot of damage, but people will find it hard to believe, but I'm seeing more damage now. It's so massive, this one. It's really sad. When I came home, back to the bayou, the next day after Ida was over, I didn't think I would have a house standing. I was surprised because my house was there. So many houses were not. The truth is that I don't know if the community will recover from this hurricane. If people do, it will be years.

Many people have similar stories after Ida. Some are moving. In the year after Ida, the Terrebonne Parish population dropped 4 percent, putting it number three in the nation for percentage of population lost between 2021 and 2022 (US Census Bureau 2023). Some are staying but must find friends or relatives to live with in other communities for the many months while their houses are being repaired. Some are in tents or RVs outside their crumpled houses. To get a perspective on the amount of damage, some places down the bayou did not get electricity service back for more than two months. Even electricity, however, can't restore the communities. A year after Ida, Our Lady of the Lake Hospital, located in lower Lafourche Parish, remained closed, except for minimal emergency services. If you stayed down the bayou, it was with the realization that you would have almost an hour drive to a hospital.

The storms are stronger, and the land is washing away more quickly than even the most-dire predictions when we started this research project (Schleifstein 2020). Many of the gardens, camps, and hunting grounds we documented during our fieldwork were destroyed or radically altered by the storm. Some people, like Wendy Billiot, are staying. She had extensive damage to her yard, but her house was in mostly good shape. She is in the process of rebuilding. Others, like Lora Ann, pondered moving. One Chauvin resident and community advocate, Jonathan Foret, said at the inaugural gathering of the Bayou Culture Collaborative, that Ida had only increased our sense of urgency and fear of cultural loss and disruption. He noted that bayou communities must find ways to transmit these valuable cultural traditions, like harvesting and sharing. "We have to teach our kids," he said, "so that they will have these traditions, no matter how far up the bayou we have to move." The key question, Foret asks, is "What would our great-grandchildren wish we had done?"[1]

✦ ✦ ✦

As this book goes to press, we are also reflecting on the COVID-19 pandemic. As millions of people were affected by stay-at-home orders in spring and summer of 2020, many who had never seriously considered it began digging up their yards to create plant beds (or creating a small garden in a patio or a sunny window). It's too soon to tell if these new gardeners and newly formed mutual aid networks will persist beyond the initial crisis. But we think those who are learning how to take care of themselves and each other through cultivating, harvesting, and sharing foods have much to learn from the residents of coastal Louisiana.

ACKNOWLEDGMENTS

Shana and Helen would like to thank Harry Luton, Shirley Fiske, Don Callaway, Melissa Poe, Alaka Wali, James Fall, and other scholars who have studied subsistence fishing and harvesting for welcoming us into the field. Our incomparable research team was composed of Wendy Wilson Billiot, Lora Ann Chaisson, Annemarie Galeucia, Mike Saunders, Jamie Digilormo, Audriana Hubbard, Jessi Parfait, Chris Adams, and Tiffany Duet. We couldn't have done it without you. We are especially grateful to all the residents of Lafourche and Terrebonne who sought us out and those who responded to requests to share their knowledge and experiences.

Colleagues who consulted with us, cheered us on, and offered sage advice include Nathalie Dajko, Gary Lafleur, Quenton Fontenot, Allyse Ferrara, Connie Sirois, Robin White, Maida Owens, Jonathan Foret, Judie Maxwell, Rachel Breunlin, David Beriss, Jeffrey Ehrenreich, Diane Austin, Brian Marks, John Protevi, Micha Rahder, Matt Keel, Alaka Wali, Glenn Thomas, Devon Turner, and Leo Gorman. Several people from Sea Grant, the LSU AgCenter, and the Louisiana Department of Wildlife and Fisheries gave us wise advice early on. Julie Falgout and Albert "Rusty" Gaudet helped us get oriented to the shrimping industry in the region.

The research was developed in partnership with the Environmental Studies Program of the Bureau of Ocean Energy Management (BOEM), Gulf of Mexico Region, and funded by the Coastal Marine Institute. Our project received the support of the Department of Geography and Anthropology and the College of Humanities and Social Science at Louisiana State University and at Nicholls State University, from the Department of English, Modern Languages, and Cultural Studies, the College of Liberal Arts, and the Center for Bayou Studies. Early versions of this work were presented at the Society for Applied Anthropology and the American Anthropological Association, the Louisiana Folklore Society, the Louisiana Academy of Sciences, Departments of Geography and Anthropology and French Studies at LSU, and Grow Dat Youth Farm in New Orleans. In particular we want to thank participants in "Rethinking Subsistence in Troubled Times: New Contexts, Configurations, and Intersections with Social and Environmental Justice," organized by Shirley Fiske at the Society for Applied Anthropology meeting in Portland for their groundbreaking research, illuminating questions, and insights. They include Donald Callaway, Syma Ebbin, Shirley Fiske, Melissa Poe, and Harry Luton. Shirley read every page of this manuscript and gave us extensive comments. Shirley and Don generously gave us permission to draw from their 2020 report. Maida Owens and Jill Brody provided edits, support, and encouragement at critical times. At the Land Memory Bank and Seed Exchange and the Neighborhood Story Project, Monique Verdin, Rachel Breunlin, and Tammy Greer gave us permission to draw from their Botanica Series.

At the University Press of Mississippi, we are especially grateful to executive editor Craig Gill; Don Davis and Carl Brasseaux, editors of the America's Third Coast Series, who encouraged us to write this book; Mary Heath, who worked with us on the home stretch; and anonymous reviewers who gave us insightful and constructive comments.

Helen would like to thank John Parker for his love and support throughout the arc of this project and the many friends who have made living in and writing about Louisiana a joy; Martha Radice, whose work on the anthropology of happiness prodded us to explore that dimension; and friends who provided support and solace and memorable meals during the period of research and writing: Antonio Garza, Michel Varisco, Nikki Thanos, Rachel Lyons, Gwen Thompkins, Rebecca Sheehan, Rachel Breunlin, and Antoinette Jackson.

Shana would like to thank the members of Humanities Write, a group of colleagues at Nicholls State University who gave up their time to read early chapter drafts: Michael Martin, Erick Piller, Robin White, Scott Banville, Alex Fabrizio Sumpter, Katie Collins, Richmond Eustis, Stephanie Baran, Shae Cox, and Allen Alexander. Other colleagues who patiently listened included Jay Udall, Tiffany Duet, Gary LaFleur, and John Doucet. Thanks also to Maida Owens and Patrick Mackin, who made invaluable contributions. Deep gratitude to Peter Jones for his photo expertise and amazing skills. A special thanks to Richard Borne for his support and all the great hallway conversations.

APPENDIX A
Discussion Questions, Resources, and Project Ideas

Some readers of this study may find that their knowledge and experience are not reflected here, and we believe much about subsistence in Louisiana remains to be studied. We want to encourage people to initiate their own documentation projects. With this in mind, we include some discussion questions to spur ideas and also some of our research tools. Food logs, oral histories, and field notes are research methods that get better with years of practice. But they are also within reach of every community. You may wish to partner with an oral historian, an anthropologist, or a social scientist at a nearby college to help you get the project off the ground. Talk to elders and document their stories. We hope this book will begin conversations and stimulate further research, especially in the many places in North America that also have strong hunting and harvesting traditions.

DISCUSSION QUESTIONS

Oyster Spaghetti: A Preface

Activity: Keep a food journal for forty-eight hours and note the foods you eat and how you eat them (raw, steamed, baked, roasted, fried, sautéed, microwaved). For each food, note

- Did you eat at home or eat out (if so where?)?
- Where does each item of food come from: Where is it grown? Where is it processed or packaged?
- Is this a food item you are familiar with from your family or growing up?
- Does this food item come with serving instructions?
- "Pin" the source of your food on a map for that forty-eight-hour period. Where do most of the pins cluster?
- What do you know about the way food is produced or harvested?
- Who produces the food on your plate?
- How does it taste?

Chapter 1: Framing Subsistence: "It's Just What We Do"

1) The authors struggled to define and explain "subsistence" to an audience that may not know the term. They ultimately drew on the metaphor of the complicated Mississippi River basin, which is full of inlets and outlets. How well does this metaphor work for you? Does it help you grasp why the definition of "subsistence" is so broad? Can you think of other concepts that are complex and multilayered like subsistence?

2) In your education, in what ways have you worked in teams? What benefits and drawbacks can you see to having a large team to work with?

3) Are there specific foods that are highly valued in your community? In your home, when you want to honor family or friends, what do you (or your relatives) cook for them?

4) What foods are the focus of holiday meals in your family tradition?

5) Does anyone in your extended family or your community grow or harvest their own food? Do any of them hunt? What stories are passed down or circulated about their food practices (gardening, fishing, hunting, etc.)?

Chapter 2: Portraits of Practice

1) Narratives about subsistence are almost as important as the food itself. Think about other activities (sports, movies, music) that you participate in where talking *about* that activity is (almost) as important as doing it. Who do you talk with and in what situation or context? What does this kind of talk say about you and your friends (or whoever you're talking to)? How does it build up a feeling of connection and being part of a group?

2) In his essay on Grand Isle, Rory Eschete writes: "The camp has everything a Cajun could need. There is a fryer, barbecue pit, smoker, boiling pot, boat, crab traps, and most importantly the twenty-by-forty-foot deck." What is the significance of this list of necessities? What does this tell you about what's important and what matters in a camp? What does it tell you about Cajun identity?

Chapter 3: Harvesting as History

1) Where does subsistence heritage entwine with the history of coastal Louisiana? Where does knowledge about the land, water, and wildlife of the region come from? How have regulations (and other restrictions to resource access) come about? How have those regulations been shaped by racial, ethnic, and class hierarchies? How were enslaved Africans involved in subsistence activities? White settlers? Immigrants? Indigenous people? How did plantation economies depend, in part, on subsistence hunting and fishing?

Chapter 4: Heritage, Identity, and Place

1) One of the central goals of this chapter is not only to describe the people who inhabit the region, but also to show how their identities are linked to the region itself. Can you think of how place has shaped identity in your own community?

2) In this chapter, the authors discuss how identities are layered and complex. This is actually common among modern people who may feel parts of multiple communities. In what ways is your own identity layered?

Chapter 5: Family, Community, and Feasting

1) How do subsistence foods (and local foods) make family gatherings possible? How do people in this chapter talk about special foods that are the draw at these gatherings?

2) In your family, what feasts or foods serve to bring families or communities together?

2) How important is taste, flavor, and the sensory qualities of food (e.g., looks, texture)?

Chapter 6: Camps, Leases, and Clubs

1) In this chapter, we see how hunting and fishing camps, hunting clubs, and leases for harvesting wild game become an important part of not only how people are able to continue hunting, but how they socialize and train children. Can you think of other social institutions that serve a similar purpose in your community? In what ways, for example, are social groups like recreational sports teams similar to or different from these camps and clubs?

2) What places are important in your family or your community for transmitting important aspects of cultural heritage or traditional practices?

Chapter 7: "Worth It" and Other Measures of Value

1) Drawing on the stories in this chapter, what kind of argument could you make that hunting is worth the financial cost? That it is not?

Chapter 8: Self-Reliance, Care, and Mutual Aid

1) Louis tells us he doesn't want his son to play baseball in the summer, "I'd like to see him be a man." What do you think of this statement? How does it suggest shrimping is linked to gender identity? Are some occupations marked as masculine or feminine in your own community? Your family? School?

2) Does it remind you of the ways that other specific occupations are gendered?

3) Are there ways that your community has of mutual aid that makes it more *tenacious*? What are the components of well-being that you see as being foundational to your family or community?

Appendix A: Discussion Questions, Resources, and Project Ideas

Chapter 9: Conclusion

To learn more about Louisiana's rapidly shrinking coastline, check out the interactive site created through a joint project of a New Orleans online news site, The Lens, and the national investigative reporting organization, ProPublica.

FILMS AND RESOURCES FOR FURTHER READING

Films

Smokin Fish by Luke Griswold-Tergis and Corry Mann (Tlingit)—Through a dry sense of humor and community oral histories, Corry Mann narrates his life story and his involvement in dual economies, salmon fishing and smoking, and sharing, and global capitalism through his mail-order business of authentic Native art objects and crafts.

Gather by Sanjay Rawal (2020), featuring Nephi Craig (White Mountain Apache), Twila Cassadore (San Carlos Apache), Elise Dubray (Cheyenne River Sioux and MHA), and Samuel Genshaw (Yurok).

My Louisiana Love by Sharon Linezo Hong with Monique Verdin (Houma). Chronicles Monique's relationship with her grandmother Matine and her Houma relatives, documenting Houma traditions, and changing landscape of coastal Louisiana. visionmakermedia.org.

Yum Yum Yum! by Les Blank / Flower films.

We Live to Eat by Kevin McCaffrey and Neil Alexander.

No One Went Hungry: Cajun Food Traditions Today by Kevin McCaffrey (vimeo).

SuperSize Me by Morgan Spurlock.

In Defense of Food with Michael Pollan.

The Gleaners and I by Agnes Varda.

Marron by Andre Gladu.

SoLa: Louisiana Water Stories by Jon Bowermaster (2010).

After the Spill: Louisiana Water Stories Part II by Jon Bowermaster (2016)—distributed by Bullfrog films and a sequel to SoLa (above).

Beasts of the Southern Wild by Court 13—a feature film that has stimulated controversy for its representation of down-the-bayou communities as an allegory for climate change and environmental racism. Poetic and evocative. A New Orleans lens on coastal Louisiana.

Come Hell or High Water: The Battle for Turkey Creek by Leah Mahan—chronicles the struggle by a teacher from Boston to protect the graves of his (formerly enslaved) ancestors bulldozed to make way for urban expansion in Gulfport, Mississippi. (2013).

By Invitation Only by Rebecca Snedeker (for a look at carnival traditions and social stratification in New Orleans).

Promised Land by Luisa Dantas (about post-Katrina rebuilding in New Orleans).

Films by the Inuit Filmmaking collective, led by Zacharias Kunuk and Norman Cohn, such as *Atanarjuat* (The Fast Runner), *Hunting with My Ancestors*, and the *Journals of Knud Rasmussen*, includes many short films documenting subsistence activities (hunting, fishing, food preparation) some of which are available for free streaming. http://www.isuma.tv/about-us.

Books

Fresh Fruit, Broken Bodies by Seth Holmes.
Eating the Ocean by Lisbeth Probyn.
Fat: The Anthropology of an Obsession by Don Kulick and Anne Meneley.
Alcohol: Social Drinking in Cultural Context by Janet Chrzan (2013—Routledge series for creative teaching).

PROJECTS

Oral History of Subsistence

Do you have subsistence heritage in your family? Interview an older relative (a grandparent, aunt, or uncle) about their (your) family stories about hunting, harvesting, and growing their own food.

Do any of your neighbors or classmates have family members who are involved in hunting, fishing, gardening, or harvesting wild foods? Do you know anyone involved in wildcrafting?

Museums

Visit an area museum to find paintings, photographs, or artifacts with links to subsistence heritage. Does this museum have baskets, nets, animal traps, or any other items on display related to home-grown or wild-harvested foods?

Folk Culture

Go to a folklife curriculum website—we recommend Louisiana Voices (louisianavoices.org)—for information on how students and community members can document their local culture. People can adapt those materials to explore local foodways and food culture. The curriculum also includes subsistence connections, such as feasting, festivals, harvest celebrations, storytelling, rituals, and more.

TRY SOME OF OUR PROJECT METHODS

Windshield surveys. On a drive make notes on each thing connected to homegrown and wild-harvested foods. Seed stores? Home improvement/sporting goods stores? (Lots of rural people travel to larger towns to buy seeds, seedlings, and hunting supplies from larger discount merchandizers.) Kitchen gardens? Homemade for-sale signs? Fishing boats? Roadside stands? What can you tell about such activities in your community from this survey?

Freezer/pantry inventories. This is a method that did not work for our community because of participant concerns about regulations. However, you may be more successful. A group of people could make note of what they have in their pantries or freezers that is homegrown or wild harvested. Fig preserves from an aunt? Deer sausage from a friend? Fish you caught on a trip? Compiling the results from the whole group may offer a clear picture of how connected people are to these practices. Purchases of hand-crafted items, such as homemade relish or jams, can provide information about aesthetics, taste, and what foods people value.

APPENDIX B
Eight Factors Used in Customary and Traditional Determinations in Alaska

Shirley Fiske and Don Callaway (2020:163) summarize much of the research on subsistence in Alaska Native households and communities in their report. In Table 6.3.2.2 they describe the eight factors that are used in Alaska to determine if communities qualify to be called subsistence communities and thus have preferential access to resources.

Eight Factors Used in Customary and Traditional Determinations in Alaska

> Factor 1: Long-term, consistent pattern of use of animal and plant species, including duration and consistency of use.
> Factor 2: Use pattern recurring in specific seasons.
> Factor 3: Methods and means of harvest, including mode of transportation, method of harvest, and composition of hunting party.
> Factor 4: Harvest near, or reasonably accessible, to residence.
> Factor 5: Handling, preparing, preserving, and storage of subsistence goods.
> Factor 6: Handing down of knowledge, including knowledge of the environment and wildlife.
> Factor 7: Distribution or sharing of resources.
> Factor 8: Reliance on a wide diversity of resources and importance of wild foods in diet.

(Fiske and Callaway 2020:163, reproduced with permission)

Shirley Fiske, when she read a draft of this book, noted the startling alignment between the Alaska research and our research in coastal Louisiana: "As you describe subsistence in your book, it seems to me that Louisiana hunters and harvesters and gardeners qualify on *every single factor* for having subsistence practices (of course no one, in our study or yours, uses that term). The term was made up for social sciences and resource management convenience in segmenting the world's economies and resource management" (Fiske 2022, personal communication).

We are grateful to Shirley Fiske and Don Callaway for permission to reproduce this here.

NOTES

OYSTER SPAGHETTI: A PREFACE

1. Louisiana is among the top ten states in the US for the highest percentage of population with diabetes. https://ldh.la.gov/assets/oph/chronic/diabetes/DiabetesFacts_LA_Final_2015.pdf.

2. Satsumas are a type of mandarin orange, similar to a clementine, originally from Japan, growing in coastal Louisiana since the late 1800s. Satsuma trees are commonly seen in south Louisiana yards.

3. Literally Oak Point, but also often spelled Pointe-au-Chien (Dog Point). Both names have been in use at least 150 years (Dajko 2020:99). Nathalie Dajko has explored the controversy over the spelling and the significance of the name in her linguistic ethnography of the role of language in place making in down-the-bayou communities of Terrebonne and Lafourche (Dajko 2020, especially pp. 96–116).

4. While much of environmental writing about coastal Louisiana identifies human-made canals as a major cause of coastal erosion and land loss, we also recognize that these are now part of the nature-cultures that make up the foundation for the mixed economies of subsistence practices. This reminds us to not romanticize these practices as somehow representing pristine survivals of ancient foodways, but rather products of the same forces that shape other parts of modern life.

1. FRAMING SUBSISTENCE: "IT'S JUST WHAT WE DO"

1. We are placing some longer citations in footnotes like this one: (Murton, Bavington, and Dokis 2016, Robbins et al. 2008, Poe et al. 2013, McClain et al. 2014, Teitlebaum and Beckley 2006, García-Quijano et al. 2015, Emery and Pierce 2005, Poe et al. 2015, Jehlička et al. 2008, Hurley et al. 2013, Hurley and Halfacre 2011; Brown et al. 1998; Brown and Toth 2001; Menzies 2010; Collings 2011; Collings et al. 2009; Fiske and Callaway 2020).

2. This assumption obviously overlooks not only the trade routes we mention but well-known histories of Indigenous people who were engaged in long-distance trade and many who generated surplus and engaged in feasting and redistribution.

3. Poverty Point is a UNESCO World Heritage site in northeast Louisiana that includes a complex of earthworks built more than 3,000 years ago, with evidence of feasting, ceremony, and trade as far north as the Great Lakes (povertypoint.us; Gibson 1994).

4. The scholarship in this area is substantial and compelling (Murton, Bavington, and Dokis 2016, Westman 2016, Hathaway 2016, Pottery 2016, Coté 2010; Li 2007; Matsutake Worlds 2009, Salmón 2012; Tsing 2015).

5. Shirley Fiske, personal communication, May 23, 2022.

6. We borrow this use of "otherwise" from Laura McTighe and Megan Raschig (2019) to "frame potentialities that are still emerging." And "to hold and open a place for relations or actions that don't quite fit into" dominant paradigms.

7. The hyphenated coauthors who publish under the pen name J. K. Gibson-Graham take the plural pronoun.

8. A totalizing logic tends to view everything from the point of view of a single concept or social force. It also tends to erase or trivialize anything (an experience or a phenomenon) that doesn't fit within that framework.

9. See the Louisiana Coastal Protection and Restoration Authority's Mid-Barataria Sediment Diversion Project, https://coastal.la.gov/project/mid-barataria-sediment-diversion/.

10. William Jankowiak, with Christina Turner, and Helen Regis, *Black Social Aid and Pleasure Clubs: Marching Associations in New Orleans*. New Orleans, LA: Jean Lafitte National Historical Park and the National Park Service, 1989.

11. Owens is the director of the Louisiana State Folklife Program. For more information, see http://www.louisianafolklife.org/LFP/main_maida_owens_bio.html.

12. We also came to understand that a history of increasing surveillance and regulation is widely felt as a loss of personal and collective food sovereignty among coastal residents. To read more about the concept of food sovereignty, distinct from notions of food security and food justice, see Holt-Giménez 2010, McRae 2016.

2. PORTRAITS OF PRACTICE

1. The Blessing of the Fleet tradition is held at the beginning of shrimp season. Community shrimping families decorate their boats, parade them down the bayou, and a priest blesses them and prays for both an abundant and safe harvest (Hubbard 2013).

2. *Boucherie*, from the French meaning a butchering session, refers to a tradition in francophone Louisiana, particularly among people who identify as Cajun, where a host invites neighbors and friends to a feast. Traditionally, the slaughtered hog is shared among those who help (Ancelet, Edwards, and Pitre 1991; Gutierrez 1992).

3. The method for drying shrimp was likely introduced by Filipino immigrants to Louisiana who sun dried the shrimp on platforms, making this a great example of a local tradition with global connections (Ho 2016).

4. South Louisiana culinary traditions are heavily influenced by Catholicism and the tradition of restricting meat consumption during Lent, the forty days leading up to Easter Sunday. Many Catholics abstain from eating meat on Fridays during Lent. And members of other faith-communities participate in this cultural tradition.

5. In one sense, a "camp" is simply a shelter near where you hunt or fish, but many families use them as a type of gathering spot or minivacation. We discuss camps more fully in chapter 6.

6. Grand Isle was hit particularly hard by Hurricane Ida in August 2021, destroying one out of every four houses on the island (Roberts 2021).

7. This question of the relationship between modernity and tradition has been an important conversation in anthropology, folklore, and related fields (see for example Bowman and Briggs 2003 and Piot 1999). In our popular culture, we tend to think of them as totally different concepts, but in fact they are closely linked. Modernity and rapid social change have caused people to become acutely self-conscious about disappearing traditions (such as languages and folkways) and to intentionally work to document, preserve, and transmit them—a process that always inevitably involves transformation.

8. "Mirliton" is the Louisiana French name for a squash more commonly called "chayote" (*Sechium edule*), a member of the squash and gourd family. This easy-to-grow, mild, green squash was originally cultivated in Mexico and came to be a Louisiana staple via Europeans who brought the squash from Central America into the port of New Orleans. People anglicize the word in two ways. Some people say "MER-luh-tan," while the older anglicization comes out pronounced something like "MILLY-tan."

3. HARVESTING AS HISTORY

1. The city is roughly one hundred miles from the mouth of the Mississippi River.

2. Medical anthropologists bring together social and biological dimensions to understand patterns as "syndemics." Recent work by Merrill Singer and Barbara Rylko-Bauer (2021:7–9) defines *syndemics* as "synergistic interactions of two or more diseases or other health conditions (e.g., nutritionally inadequate diet) [linked to] social and environmental conditions" (7–8) and *structural violence* as "the often-hidden ways that structures of inequality, such as poverty, racism, and discrimination, negatively impact the lives and well-being of affected populations" (8). "Structural violence drives syndemics" (8) and helps to explain "the clustering of multiple diseases and risks in vulnerable populations" (Singer and Rylko-Bauer 2021:9).

3. The Centers for Disease Control consider people "overweight" if their BMI is 25 or more, and "obese" if their BMI is 30 or more.

4. We use the terms Native American, Indigenous, Native peoples, and American Indian interchangeably throughout this text. All of these terms are in use in this region, varying from person to person and based on situation. For individuals, we try to use the term they prefer, which is often to note their specific tribal affiliation. For more on the shifting terminology, see Dajko 2020:159n5.

5. Other groups also settled along Bayou Lafourche, and the Isleños, Spanish-speaking immigrants from the Canary Islands, were resettled to the bayou earlier than the Acadians (Din 1999). However, by the 1800s, the Acadians were the dominant group along the bayou, particularly in Lafourche Parish.

6. For example, at the beginning of the sugar boom in the 1820s, there were only a few Anglophone plantation owners. By the start of the Civil War, there were three times as many Anglo Americans as Francophones operating plantations just in Terrebonne Parish (Cenac 2017).

7. Louisiana had many maroon communities. The best documented were north and east of the Terrebonne-Lafourche. In fact, historians believe that for a time in the 1770s maroon communities controlled the *Bas du Fleuve*, territory from the mouth of the Mississippi up to New Orleans (Whitney Plantation, n.d.), including a fairly large settled community, St. Malo near Lake Borgne in what is now St. Bernard Parish (Diouf 2014). St. Malo, which was

4. HERITAGE, IDENTITY, AND PLACE

1. Lora Ann explains: "This is the white jambalaya that I learned from my mom. That she had learned from a friend she would play cards with. All of us make a red jambalaya and a brown jambalaya, not too many people make this particular jambalaya." She calls for one can each of cream of celery soup, French onion soup, and Rotel tomatoes, a brand of canned tomatoes that include bits of jalapeño chilies. Local cooks use a variety of herbs, like thyme, oregano, and bay leaf, some cayenne (to give it some heat), and hot sauce to taste at the table.

2. Most of these people were in Bourg or Chauvin. Bourg is located along Bayou Terrebonne about a fifteen-minute drive south of Houma. Ironically, both Bourg and Chauvin are no more than fifteen minutes north of places such as Pointe-aux-Chênes and Dulac, which have large Native American communities (see Dajko 2020).

3. Although awareness has grown, we remind you of Monique Verdin's words (quoted in chapter 3) about the forgotten links between place names and Indigenous people: "You know, I have to tell people sometimes that. They're like, 'Oh! Houma?!' [. . .] 'Oh, there's Indians there?'" (Breunlin and Verdin 2020).

4. Dr. Greer was speaking on October 13, 2020, as a guest in the Botanica Series organized by Monique Verdin and Rachel Breunlin and hosted by the Neighborhood Story Project and the Land Memory Bank and Seed Exchange. This talk was one of many programs supported by the Louisiana Folklore Society's Bayou Culture Collaborative, which is a partnership with the Louisiana Division of the Arts Folklife Program and the South Louisiana Wetlands Discovery Center.

5. Tidwell transcribed this quote in a way he hoped captured Cajun English, specifically the way some people who grew up speaking French change the "th" sound to either "d" or "t." So "then" becomes "den," and "with" becomes "wit." However, we are presenting it in a more customary writing style. The original quote was written as "For me, it's when dat old mornin' sun comes risin' over my boat deck and de boat's covered wit' a ton of shrimp, and den I have me a big bowl of jambalaya wit' de guys at de shed, and we all ever'body got money in de pocket. Dat, to me, dat's bein' Cajun!" (Tidwell 2004:57). The "shed" refers to the shrimp shed, where shrimpers unload or sell their catches.

6. "Lagniappe" is a Louisiana French word, usually pronounced "lan-yap," which means giving someone a little something extra. For example, "I went to my neighbor's house to pick satsumas, and she gave me a cabbage as lagniappe."

7. Shirley Fiske, personal communication, May 23, 2022.

8. In personal communications and in class discussions about identifying as Cajun, Shana has talked to residents, as well as some students at Nicholls, who identify as both African American and Cajun. Most, however, do not. In contrast, some African American students have expressed frustration with depictions of Terrebonne and Lafourche as "Cajun Country."

9. One student told Shana that he considered his relatives to be Cajun but he wasn't because he was only interested in computers.

10. Ms. Bennett made this remark during a discussion at a 2019 gathering sponsored by the South Louisiana Wetlands Discovery Center in Chauvin.

5. FAMILY, COMMUNITY, AND FEASTS

1. Crabs can be legally caught year-round in Louisiana, although some years there are restrictions on whether commercial crabbers can harvest female blue crabs in March and April. However, Jacob is not referring to an enforced legal season but rather an informal season, usually from April through July, months when harvesting blue crab is at its peak.

2. Morgan City is a coastal town in St. Mary Parish, about forty minutes west of Thibodaux, located at the mouth of the Atchafalaya River. It hosts the Shrimp and Petroleum Festival to celebrate the region's bounty. Flat Lake, where the crabber drops his traps, is located a few miles outside of Morgan City, one of a series of interconnected natural lakes in the Atchafalaya Basin.

3. Shirley Fiske told us that she and Don Callaway documented this pattern among DC area subsistence fishers as well. They write, "Among African Americans, 89% learned to fish from family members, including great-grandmothers in the rural south and fourth generation Washingtonian fathers in D.C. One fisherman told us he learned from his great-grandmother in South Carolina, and thought her great-grandparents had taught her, harkening back to a more subsistence lifeway in earlier generations" (Fiske and Callaway 2020:172).

4. Black drum, *Pogonias cromis*, is a common fish caught in the region, often found in brackish waters.

5. We used Voyant, a web-based analysis for digital texts that provides limited corpus analysis, including word frequency counts. The data was filtered for stop words using the TAPoRware list prior to processing.

6. This reality show premiered in 2010 on the History Channel and was about to launch its thirteenth season in 2022. The show has featured multiple crews of alligator hunters, often portrayed in a humorous, but stereotypical fashion.

7. A 20-gauge shotgun, a smaller, lighter gun with less recoil, is often considered a first choice for a youth gun. The "crack barrel" (also called "break barrel") means that only one shot can be loaded at a time. Often the stocks are trimmed down to enable children to hold the gun better.

8. The season for hunting migratory game fowl, like ducks and geese, is governed in part by international treaties and federal regulations. State officials work with federal officials to set the number of days hunters can shoot, often splitting the days into two or three sets of days, rather than one solid season. This has prevented waterfowl in the US from being over hunted. The effort to have continental regulation of waterfowl was initiated by hunters, particularly those, like John Serigny, involved with Ducks Unlimited (ducksunlimited.org).

9. Suffering through a first experience showed up in many stories we recorded about initiation into hunting and fishing. Our narrators often specified that they brought few supplies, they had little or no prepared food, and they had to rely on what they caught or killed. That experience of scarcity is key. And it foreshadows the abundance that comes with skill and experience. Some told us of being dropped off on a levee with nothing but their rifles. Others remembered going out on the boat with few supplies, having to rely on what they caught. Store-bought supplies (oil, cornmeal, or milk) are often named explicitly, as if to demonstrate how few were carried. Sometimes they name what they wish they had brought (a bologna sandwich) to underline their hunger with humor. The feeling is that hard lessons learned from hunger are key to being initiated into this way of life.

6. CAMPS, LEASES, AND CLUBS

1. Brule is pronounced **brew**-*lee*, and Labadieville is pronounced **lab**-*uh-dee-ville*. Both have stress on the first syllable. Labadieville sits on Bayou Lafourche, about ten miles west of Thibodaux, going up the bayou toward its source, the Mississippi River.

2. Nutria are semiaquatic rodents native to South America and introduced to the US in the 1930s as a fur-bearing species. A lack of predators (due to human intervention), rapid breeding rates, and the expansive habitat offered by southern Louisiana have led to their overpopulation. Destructive of levees and riparian habitat, they are considered a nuisance animal and are not widely eaten. The state will pay hunters a $5 bounty for each nutria they kill.

3. A "blind" is simply a shelter to hide you from deer, ducks, or whatever you are hunting. In this case, the hunters at the camp had built platforms in the trees where they were hidden from the deer and had a better view to shoot.

4. Blue tongue disease is a virus, transmitted by flies and gnats, that can be deadly for deer.

5. Richard was allowed to bring a guest to hunt, which is how a member of our research team, Mike, was able to participate.

7. "WORTH IT" AND OTHER MEASURES OF VALUE

1. Poe et al. use the phrase "subsistence share systems" and "community share systems," which we echo here. We prefer to frame subsistence as a practice (or a set of practices) integrated into everyday life, but that doesn't always look like a "system" (Poe et al. 2015:248). For a useful view of harvesting as a practice, see Robbins et al. (2008).

2. The program had a problem with two-word entries. The most obvious problems are foods like "poule d'eau" and "green beans," splitting them into two. Another problem is the use of two words for the same thing, like "deer" and "venison." Nevertheless, the cloud captures both the breadth and relative frequency of harvested foods. A poule d'eau is a water bird, commonly called a coot in other parts of the US.

3. In Louisiana courtbouillon is a hearty fish stew, usually with a roux base and tomato sauce, sometimes spelled "coubion." See Harris 2003.

4. Here "Japan plum" refers to the loquat plant, *Eriobtrya japonica*, also known in the region as "misbelieve."

5. Houses in the lower parts of Terrebonne and Lafourche Parish are often raised, sometimes by as much as ten or twelve feet from the ground because of frequent flooding. This under-house space gets used for various purposes. Some people park cars or boats there or have picnic tables and children's play areas. Some people use the shaded space to process or prepare harvested foods—plucking ducks or filleting fish.

6. Shirley Fiske and Don Callaway found a similar emphasis on sharing among subsistence fishers in their research in the DC area. "We found [sharing] was a *huge* part of the fishing ethos in D.C." They write: "Almost everyone, 97% of the fishers, share their harvest, and 18% are 'share only,' meaning that they do not personally consume the fish, but share or give away what they catch" (Fiske and Callaway 2020:186). That's virtually 100 percent of their interviewees. They also asked fishers "with whom do you share?" They found a similar pattern as in Louisiana. "People have priorities in their sharing. Most of the fishermen share

with immediate family or kin first, and then friends and neighbors" (Shirley Fiske, personal communication, May 23, 2022; Fiske and Callaway 2020:186).

7. Sharon Hutchinson's classic study of the economic transformation among the Nuer of East Africa finds that hybrid categories (cattle of money, cattle of girls) help to mediate social change and smuggle social values and relationships into cash exchanges (1992).

8. Explanations for some items in this list: Filé is a spice made from the dried and ground leaves of a sassafras tree. Commonly used in gumbo. Choupic (*Amia calva*) is often spelled choupique (but not on this sign) and is commonly called a bowfin or swamp trout in other parts of the US. As noted in the preface, satsumas are a type of mandarin orange, similar to a clementine. A cushaw squash, also called a cushaw pumpkin, silver-seed gourd, green-striped cushaw, or juirdmon (in Louisiana French), is a large crookneck squash (they can weigh up to twenty pounds) that has a long history in south Louisiana, particularly used to make cushaw pies (Elie 2009). "Field peas" can be a catch-all term for several varieties of legume. Commonly grown varieties include crowder, purple hull, and pinkeye. The only variety commonly available in stores is the black-eyed pea. These peas are grown in gardens throughout the US South.

8. SELF-RELIANCE, CARE, AND MUTUAL AID

1. Verdin emphasizes that "Botanica is a multi-racial/ethnic collaboration that pulls together storytellers, scholars, herbalists, museums, artists, and gardeners to cross-pollinate knowledge of ethnobotany across communities in south Louisiana. At the heart of this project is the idea of reciprocal sharing within bayou communities—introducing traditional Indigenous knowledge of plants that Houma communities have preserved in their gardens and inviting healers in other communities to share their healing work with plants and gardens. It is our hope that this project will create bridges between communities who have been segregated from one another to create long-lasting relationships. The work will include oral histories, portraits, photographs of plants and their healing properties, and recipes" (personal communication, August 26, 2022).

2. A "dos gris" is a local term for a type of duck more commonly called "bluebill" in other parts of the US or "scaup" in some parts of the US and Europe. The scientific name is *Aythya marila*. Like dos gris (or scaup), pintail ducks, *Anas acuta*, are found throughout North America.

3. The Louisiana Department of Wildlife and Fisheries sets the opening of the spring shrimping season, which runs roughly from May through July. The fall white shrimp season is usually open from mid-August until mid-December.

4. (Chudakova 2017; Mattingly 2014; Mol et al. 2010; Puig de la Bellacasa 2015).

9. CONCLUSION

1. A recent volume bringing together indigenous scholars, activists, cultural workers, and scientists makes connections explicit: *Indigenous Food Sovereignty in the United States: Restoring Cultural Knowledge, Protecting Environments, and Regaining Health*, edited by Devon A. Mihesuah and Elizabeth Hoover (Norman, OK: University of Oklahoma Press, 2019).

2. There is a vibrant debate about how the term *resilience* has been used by scholars, nonprofits, activists, and public agencies (see for example Barrios 2016, Browne 2016, Clark 2019, MacKinnon and Derickson 2013). The difficulty comes with the use of the term to evaluate communities, as if the capacity to recover were inherent in a culture or a people, rather than being linked to systems and structures. At worst, the term is used to "blame the victim" after a disaster for their failures of resilience (Schuller 2012, 2016). Some coastal activists choose the word *tenacious*. One bayou person explained that, to him, being resilient was just about surviving what happens to him, whereas being tenacious is about "my own willingness to carry on, my own abilities."

3. Jonathan Foret used this expression in his talk "Sinking in: Culture and Disruption," Bayou Culture Gathering, January 28, 2022. The talk is archived by the Louisiana Folklore Society at https://youtu.be/nhXaIkYmhxk.

4. Her book is part of a genre of studies seeking to "understand" the political consciousness of white working-class men, which risks framing them as a "problem" to be solved. As scholars who live and work in the region, we are saddened and frustrated by another example of local realities and lived experiences, in all their complexity, eluding national experts.

5. It may be useful to talk of these practices in terms of precarity, assemblages, and world-making, as does Anna Tsing in her 2015 book *The Mushroom at the End of the World*. For Tsing, world-making "focuses on practical activities rather than cosmologies" (or philosophies of being). For Tsing, "every organism makes worlds, humans have no special status. Finally world-making projects overlap. . . . [T]hinking through world-making allows layering and historically consequential frictions" (Tsing 2015:292n7).

POSTSCRIPT: HURRICANE IDA

1. See Foret's 2022 talk, "Sinking In: Culture and Disruption" on the Louisiana Folklore Society YouTube Channel, https://www.youtube.com/@louisianafolkloresociety2277.

REFERENCES

Aidells, Bruce. 2008. "Corn Maque Choux." *Bon Appetit*, October. Digital resource. Accessed May 13, 2019. https://www.epicurious.com/recipes/food/views/corn-maque-choux-350113.

Ancelet, Barry Jean, Jay Edwards, and Glen Pitre. 1991. *Cajun Country*. Jackson, MS: University Press of Mississippi.

Anderson, Elizabeth. 1995. *Value in Ethics and Economics*. Cambridge, MA: Harvard University Press.

Austin, Diane. 2014. "Guestworkers in the Fabrication and Shipbuilding Industry along the Gulf of Mexico: An Anomaly or a New Source of Labor?" In *(Mis)managing Migration: Guestworkers' Experiences with North American Labor Markets*, edited by David Griffith. Santa Fe, NM: SAR Press.

Austin, Diane. 2006. "Coastal Exploitation, Land Loss, and Hurricanes: A Recipe for Disaster." *American Anthropologist* 108(4):671–91.

Austin, Diane E. 2003. "Community-Based Collaborative Team Ethnography: A Community-University-Agency Partnership." *Human Organization* 62(2):143–52.

Austin, Diane E. 2004. "Partnerships, Not Projects! Improving the Environment through Collaborative Research and Action." *Human Organization* 63(4):419–30.

Austin, Diane, Brian Marks, Kelly McClain, Tom McGuire, Ben McMahan, Victoria Phaneuf, Preetam Prakash, Bethany Rogers, Carolyn Ware, and Justina Whalen. 2014. *Offshore Oil and the Deepwater Horizon: Social Effects on Gulf Coast Communities*. Volume I: *Methodology, Timeline, Context, and Communities*. OCS Study. BOEM 2014-617. US Department of the Interior, Bureau of Ocean Energy Management, Gulf of Mexico OCS Region.

Austin, Diane, Shanon Dosemagen, Brian Marks, Tom McGuire, Preetam Prakash, and Bethany Rogers. 2014. *Offshore Oil and the Deepwater Horizon: Social Effects on Gulf Coast Communities*. Volume II: *Key Economic Sectors, NGOs, and Ethnic Groups*. OCS Study. BOEM 2014-618. US Department of the Interior, Bureau of Ocean Energy Management, Gulf of Mexico OCS Region.

Barnes, Bruce "Sunpie," and Rachel Breunlin. 2016. "Pasajs: Passages for San Malo." *South Writ Large: Stories, Arts, and Ideas from the Global South*. (Spring 2017) Southwritlarge.com.

Barra, M. P. 2021. "Good Sediment: Race and Restoration in Coastal Louisiana." *Annals of the American Association of Geographers* 111(1):266–82.

Bauman, Richard, and Charles Briggs. 2003. *Voices of Modernity: Language Ideology and the Politics of Inequality*. Cambridge, UK: Cambridge University Press.

Bavington, Dean, and Jennifer Hough Evans. 2016. "Research by People: A Panel Discussion on Living Subsistence Locally." Pp. 295–317 in *Subsistence under Capitalism: Historical and Contemporary Perspectives*, edited by James Murton, Dean Bavington, and Carly Dokis. Montreal, CA: McGill-Queen's University Press.

Bazet, R. A. 1934. *Souvenir of Centennial Celebration, Houma, Louisiana, May 10–13, 1934*. Morgan City, LA: King-Hannaford.

Belic, Roko. 2011. *Happy*. Shady Acres / Wadi Rum Films (via Netflix).

Bell, Ellen Baker. 2021. Thibodaux Massacre. *64 Parishes*. Digital resource. https://64parishes.org/entry/thibodaux-massacre.

Berkes, Fikret. 1988. "Subsistence Fishing in Canada: A Note on Terminology." *Arctic* 41(4):319–20.

Berkes, Fikret. 1990. "Native Subsistence Fisheries: A Synthesis of Harvest Studies in Canada." *Arctic* 43(1):35–42.

Bernard, Shane. 2011. "Cajuns." *64 Parishes*, a magazine of the Louisiana Endowment for the Humanities. First published July 2011. https://64parishes.org.

Bilby, Kenneth. 2008. *True-Born Maroons*. Gainesville, FL: University Press of Florida.

Blank, Les, and Maureen Gosling. 1990. *Yum Yum Yum! A Taste of Cajun and Creole Cooking*. El Cerrito, CA: Flower Films.

Boglioli, Marc. 2009a. *A Matter of Life and Death: Hunting in Contemporary Vermont*. Amherst, MA: University of Massachusetts Press.

Boglioli, Marc. 2009b. "Illegitimate Killers: The Symbolic Ecology and Cultural Politics of Coyote-Hunting Tournaments in Addison County, Vermont." *Anthropology and Humanism* 34(2):203–18.

Bourdieu, Pierre. 1977. *Outline of a Theory of Practice*. Cambridge, UK: Cambridge University Press.

Branch, G. M., M. Hauck, N. Sigwana-Ndulo, and A. H. Dye. 2002. "Defining Fishers in the South African Context: Subsistence, Artisanal, and Small-Scale Commercial Sectors." *South African Journal of Marine Science* 24:475–87.

Brasseaux, Carl. 1987. *The Founding of New Acadia: The Beginnings of Acadian Life in Louisiana, 1765–1803*. Baton Rouge, LA: Louisiana State University Press.

Brasseaux, Carl. 1992. *Acadian to Cajun: Transformation of a People, 1803–1877*. Jackson, MS: University Press of Mississippi.

Brown, Jabari, Kevin Connell, Jeanne Firth, and Theo Hilton. 2020. "The History of the Land: A Relational and Place-Based Approach toward (More) Radical Food Geographies." *Human Geography* 13(3):242–52.

Brown, Ralph B., and John F. Toth. 2001. "Natural Resource Access and Interracial Associations: Black and White Subsistence Fishing in the Mississippi Delta." *Southern Rural Sociology* 17:81–110.

Brown, Ralph B., Xiaohe Xu, and John F. Toth Jr. 1998. "Lifestyle Options and Economic Strategies: Subsistence Activities in the Mississippi Delta." *Rural Sociology* 63(4):599–623.

Browne, Katherine E. 1995. "Who Does and Who Doesn't Earn 'Off the Books'? The Logic of Informal Economic Activity in Martinique." *Anthropology of Work Review* 16(1 & 2):23–33.

Browne, Katherine E. 2015. *Standing in the Need: Comfort, and Coming Home after Katrina*. Austin, TX: University of Texas Press.

Browne, Katherine E. 2016. "Roux and Resilience: Eleven Years after Hurricane Katrina." *Sapiens*. August 31, 2016. https://www.sapiens.org/culture/hurricane-katrina-aftermath-roux-resilience/.

Bruner, Jerome. 2001. "Self-making and World-making." Pp. 25–38 in *Narrative and Identity: Studies in Autobiography, Self, and Culture*, edited by Jens Brockmeier and Donal A. Carbaugh. Amsterdam: John Benjamins Publishing.

Butler, Caroline F., and Charles Menzies. 2007. "Traditional Indigenous Knowledge and Indigenous Tourism." Pp. 15–27 in *Tourism and Indigenous Peoples*, edited by R. Butler and T. Hinch. San Francisco, CA: Elsevier.

Campanella, Richard. 2008. *Bienville's Dilemma: A Historical Geography of Louisiana*. Lafayette, LA: University of Louisiana at Lafayette Press.

Caroll, Matthew S., Keith Allan Blatner, and Patricia J. Cohn. 2003. "Somewhere Between: Social Embeddedness and the Spectrum of Wild Edible Huckleberry Harvest and Use." *Rural Sociology* 68:319–42.

Cenac, Christopher. 2017. *Hard Scrabble to Hallelujah, Volume 1: Bayou Terrebonne: Legacies of Terrebonne Parish, Louisiana*. Jackson, MS: University Press of Mississippi.

Cenac, Christopher, Clifton Theriot, and Claire Domangue Joller. 2013. *Livestock Brands and Marks: An Unexpected Bayou Country History: 1822–1946 Pioneer Families: Terrebonne Parish, Louisiana*. Jackson, MS: University Press of Mississippi.

Centers for Disease Control (CDC). National Diabetes Statistics Report. 2020. Atlanta, GA: Centers for Disease Control and Prevention, US Department of Health and Human Services.

Cherry, Rachel E. 2015. *Forgotten Houma*. Charleston, SC: Arcadia Publishing.

Chudakova, Tatiana. 2017. "Caring for Strangers: Aging, Traditional Medicine, and Collective Self-Care in Post-Socialist Russia." *Medical Anthropology Quarterly* 31(1):78–96.

Clark, John P. 2019. "Against Resilience: Hurricane Katrina and the Politics of Disavowal." Pp. 72–98 in *Between Earth and Empire: From the Necrocene to the Beloved Community*. Oakland, CA: PM Press.

Collings, Peter. 2011. "Economic Strategies, Community, and Food Networks in Ulukhaktok, Northwest Territories, Canada." *Arctic* 64(2):207–19.

Collings, Peter, George Wenzel, and Richard G. Condon. 2009. "Modern Food Sharing Networks and Community Integration in the Central Canadian Arctic." *Arctic* 51(4):301–14.

Collins, Jane. 1995. "Multiple Sources of Livelihood and Alternative View of Work: Concluding Remarks." *Anthropology of Work Review* 16(1 & 2):43–46.

Community Economies Collective. 2019. "Community Economy." Pp. 56–63 in *Keywords in Radical Geography: Antipode at 50*, edited by the Antipode Editorial Collective. London, UK: John Wiley & Sons.

Condon, Richard, Peter Collings, and George Wenzel. 1995. "The Best Part of Life: Subsistence Hunting, Ethnicity and Economic Adaptation among Young Adult Inuit Males." *Arctic* 48(1):31–46.

Coté, Charlotte. 2010. *Spirits of Our Whaling Ancestors: Revitalizing Makah and Nuu-chah-nulth Traditions*. Seattle/Vancouver: University of Washington / UBC Press.

Cox, Shae, and Kevin McQueeney. 2022. "The History of Smithridge: Oral Histories." Nicholls State University Scholars Expeaux, April 14, 2022.

Dajko, Natalie. 2020. *French on Shifting Ground. Cultural and Coastal Erosion in Coastal Louisiana*. Jackson, MS: University Press of Mississippi.

Dajko, Natalie. 2019. "History and Variation in Louisiana French." Pp. 75–89 in *Language in Louisiana: Community and Culture*, edited by Natalie Dajko and Shana Walton. Jackson, MS: University Press of Mississippi.

Dardar, T. Mayheart. 2007. *Pointe Ouiski: The Houma People of LaFourche-Terrebonne*. Houma, LA: United Houma Nation.

Darensbourg, Jeffery U., ed. 2018. *Bulbancha Is Still a Place: Indigenous Culture from New Orleans*. Number 1: *The Tricentennial Issue*, September 2018. Available at: http://bulbanchaisstillaplace.org/.

Darensbourg, Jeffery U., ed. 2019. *Bulbancha Is Still a Place: Indigenous Culture from "New Orleans."* Number 2: *The Language Issue*, April 2019.

Davis, Don. 2010. *Washed Away? The Invisible Peoples of Louisiana's Wetlands*. Lafayette, LA: University of Louisiana at Lafayette Press.

Davis, Don. 1973. Louisiana Canals and Their Influence on Wetland Development. PhD dissertation, Department of Geography and Anthropology, Louisiana State University, Baton Rouge.

Dawdy, Shannon Lee. 2010. "'A Wild Taste': Food and Colonialism in Eighteenth-Century Louisiana." *Ethnohistory* 57(3):389–414.

DeSantis, John. 2016. *The Thibodaux Massacre: Racial Violence and the 1887 Sugar Cane Labor Strike*. Charleston, SC: History Press.

Diouf, Sylviane. 2014. *Slavery's Exiles: The Story of American Maroons*. New York, NY: New York University Press.

Ditto, Tanya Brady. 1980. *The Longest Street: A Story of Lafourche Parish and Grand Isle*. Baton Rouge, LA: Moran Publishing.

Dochuk, Darren. 2019. *Anointed with Oil: How Christianity and Crude Made Modern America*. New York, NY: Basic Books.

Duffy, McFadden. 1969. *Hunting and Fishing in Louisiana*. New Orleans, LA: Pelican Publishing House.

Duncan, Colin A. M. 2016. "On the Semantics of Theorizing the Cause(s) of the Shadows, or How to Think about Counting the Differences between a Wild Edible Mushroom and a Super Tanker, Neither of Which Fits the Commodity Form." Pp. 346–67 in *Subsistence under Capitalism: Historical and Contemporary Perspectives*, edited by James Murton, Dean Bavington, and Carly Dokis. Montreal, CA: McGill-Queen's University Press.

Eckert, Penelope, and Sally McConnell-Ginet. 1992a. "Communities of Practice: Where Language, Gender and Power All Live." Pp. 89–99 in *Locating Power: Proceedings of the 1992 Berkeley Women and Language Conference*, edited by Kirin Hall, Mary Buchotz, and Birch Moonwomon. Berkeley, CA: Berkeley Women and Language Group.

Eckert, Penelope, and Sally McConnell-Ginet. 1992b. "Think Practically and Look Locally: Language and Gender as Community-Based Practice." *Annual Review of Anthropology* 21:461–88.

Elie, Lolis Eric. 2006. "Good Gourd Almighty." *New Orleans Times Picayune*. November 16.

Ellzey, Bill. 2006. "Terrebonne Trappers Were Tops in the Nation in 1925." *Courier*. Houma, LA. December 13. Available at: http://www.houmatoday.com/article/20061213/FEATURES/612130302.

Emery, Marla, and Alan R. Pierce. 2005. "Interrupting the Telos: Locating Subsistence in Contemporary US Forests." *Environment and Planning A* 37(6):981–93.

Erickson, Kenneth Cleland, and Donald Stull. 1997. *Doing Team Ethnography: Warnings and Advice*. New York, NY: Sage Publications (digital edition: Sage Research Methods).

Fischer, Edward F. 2014. *The Good Life: Aspiration, Dignity, and the Anthropology of Wellbeing*. Stanford, CA: Stanford University Press.

Fisher, Edward F. 2017. "Beyond Nutrition: Eating, Innovation, and Cultures of Possibility." *Sight and Life* 31(1):32–39.

Fiske, Shirley, and Don Callaway. 2019. "'Fishing for Food': Subsistence Fishing in Urban Rivers and Implications for Environmental Justice." Paper presented at the Annual Meeting of the Society for Applied Anthropology, Portland, Oregon, March 22, 2019.

Fiske, Shirley, and Don Callaway. 2020. *Ethnographic Resource Study. Subsistence Fishing on the Potomac and Anacostia Rivers*. Final Report. 369 pages. Cooperative agreement between the University of Maryland and the National Park Service, National Capital Region. Washington, DC: National Park Service.

Foret, Jonathan 2022 "Sinking In: Culture and Disruption." Bayou Culture Gathering, Louisiana Folklore Society, January 28, 2022, digital resource: https://youtu.be/nhXaIkYmhxk.

Freeman, Milton M. R. 1993. "The International Whaling Commission, Small-Type Whaling, and Coming to Terms with Subsistence." *Human Organization* 52(3):243–51.

Garcia-Quijano, C., John J. Poggie, Anna Pitchon, and Miguel H. Del Pozo. 2015. "Coastal Resource Foraging, Life Satisfaction and Well Being in Southeastern Puerto Rico." *Journal of Anthropological Research* 71:145–67.

Gayarré, Charles. 1851. *Louisiana: Its Colonial History and Romance*. Manhattan, NY: Harper & Brothers.

Gibson, Jon. 1994. "Over the Mountain and across the Sea: Regional Poverty Point Exchange." *Louisiana Archaeology* 17:251–99.

Gibson-Graham, J. K. 2006. *The End of Capitalism (as We Knew It): A Feminist Critique of Political Economy*. London, UK: Blackwell.

Gibson-Graham, J. K. 2008. "Diverse Economies: Performative Practices for 'Other Worlds.'" *Progress in Human Geography* 32(5):613–32.

Giltner, Scott E. 2008. *Hunting and Fishing in the New South: Black Labor and White Leisure after the Civil War*. Baltimore, MD: Johns Hopkins University Press.

Gonzales, Randy. 2019. "Stories Told about the 19th Century Filipino Settlement at St. Malo, Louisiana." *Louisiana Folklife Miscellany* 29:5–22.

Gowland, Bryan. 2003. "The Delacroix Isleños and the Trappers' War in St. Bernard Parish." *Louisiana History* 44(4):411–41.

Gramling, Robert, and Ronald Hagelman. 2005. "A Working Coast: People in the Louisiana Wetlands." *Journal of Coastal Research* SI(44):111–33.

Greer, Tammy. 2020. "A Talk with Dr. Tammy Greer." *Botanica: A Series of Conversations*. New Orleans, LA: Land Memory Bank and Seed Exchange in partnership with the Neighborhood Story Project, October 13, 2020.

Greer, Tammy. 2015. "Medicine Wheel Garden." *Medicine Wheel Garden Videos* 1. https://aquila.usm.edu/garden_videos/1.

Griffith, David, and Manuel Valdéz Pizzinni. 2002. *Fishers at Work, Workers at Sea: A Puerto Rican Journey through Labor and Refuge*. Philadelphia, PA: Temple University Press.

Guidry, Sherwin. 1970. *Le Terrebonne: A History of Montegut*. Montegut, LA: published by author.

Gutierrez, Paige C. 1992. *Cajun Foodways*. Jackson, MS: University Press of Mississippi.

Hall, Gwendolyn Midlo. 1995. *Africans in Colonial Louisiana: The Development of Afro-Creole Culture in the Eighteenth Century*. Baton Rouge, LA: Louisiana State University Press.

Hall, Gwendolyn Midlo. 1992. *African in Colonial Louisiana: The Development of Afro-Creole Culture in the Eighteenth Century*. Baton Rouge, LA: Louisiana State University Press.

Hathaway, Michael J. 2016. "Rethinking the Legacies of Subsistence Thinking." Pp. 234–53 in *Subsistence under Capitalism: Historical and Contemporary Perspectives*, edited by James Murton, Dean Bavington, and Carly Dokis. Montreal, CA: McGill-Queen's University Press.

Ho, Winston. 2016. "Researching Chinese American History in New Orleans: Manila Village." (posted July 1, 2016, revised 2018) Available at: https://nolachinese.wordpress.com/2016/07/01/manila-village/.

Hochschild, Arlie Russell. 2016. *Strangers in Their Own Land: Anger and Mourning in the American Right*. New York, NY: New Press.

Holt-Giménez, Eric. 2010. "Food Security, Food Justice, or Food Sovereignty?" *Food First Backgrounder*: Institute for Food and Development Policy 16(4):1–4.

Holt-Giménez, Eric, and Annie Shattuck. 2011. "Food Crises, Food Regimes, and Food Movements: Rumblings of Reform or Tides of Transformation?" *Journal of Peasant Studies* 38(1):109–44.

Hotard, Corey David. 2015. Just Throw It in the Pot! The Cultural Geography of Hidden Landscapes and Masked Performances in South Louisiana Gumbo Cooking. PhD dissertation, Department of Geography and Anthropology, Louisiana State University, Baton Rouge.

Hubbard, Audriana. 2013. The Blessing of the Fleet: Heritage and Identity in Three Gulf Coast Communities. MA thesis, Department of Geography and Anthropology, Louisiana State University, Baton Rouge.

Hunt, Alfred N. 1988. *Haiti's Influence on Antebellum America: Slumbering Volcano in the Caribbean*. Baton Rouge, LA: Louisiana State University Press.

Hurley, Patrick, and Angela Halfacre. 2011. "Dodging Alligators, Rattlesnakes, and Backyard Docks: A Political Ecology of Sweetgrass Basket-Making and Conservation in the South Carolina Lowcountry, USA." *GeoJournal* 76:383–99.

Hutchinson, Sharon. 1992. "The Cattle of Money and the Cattle of Girls among the Nuer, 1930–1983." *American Ethnologist* 19(2):294–316.

Illich, Ivan. 1973. *Tools for Conviviality*, London, UK: Marion Boyars.

Illich, Ivan. 1981. *Shadow Work*. London, UK: Marion Boyars.

Jackson, Antoinette T. 2012. *Speaking for the Enslaved: Heritage Interpretation at Antebellum Plantation Sites*. Walnut Creek, CA: Left Coast Press.

Jackson, Antoinette T. 2020. *Heritage, Tourism, and Race: The Other Side of Leisure*. New York, NY: Routledge.

Jacob, Michelle M. 2013. *Yakama Rising: Indigenous Revitalization, Activism, and Healing*. Tucson, AZ: University of Arizona Press.

Jankowiak, William, with Christina Turner, and Helen Regis. 1989. *Black Social Aid and Pleasure Clubs: Marching Associations in New Orleans*. New Orleans, LA: Jean Lafitte National Historical Park and the National Park Service.

Jehlička, Petr, Tomáš Kostelecky, and Joe Smith. 2008. "Food Self-Provisioning in Czechia: Beyond Coping Strategy of the Poor: A Response to Alber and Kohler's 'Informal Food Production in the Enlarged European Union.'" *Social Indicators Research* 111:219–34.

Johnson Gaithere, Cassandra, Amada Aragón, Marguerite Madden, Sheridan Alford, Aza Wynn, and Marla Emery. 2020. "'Black Folks *Do* Forage': Examining Wild Food Gathering in Southeast Atlanta Communities." *Urban Forestry and Urban Greening* 56:1–9.

Johnston, Barbara Rose. 2012. "On Happiness. Vital Topics Forum." *American Anthropologist* 114(1):6–18.

Kidder, Tristram R. 2000. "Making the City Inevitable: Native Americans and the Geography of New Orleans." Pp. 9–21 in *Transforming New Orleans and Its Environs: Centuries of Change*, edited by Craig Colten. Pittsburgh, PA: University of Pittsburgh Press.

Kimmerer, Robin Wall. 2015. *Braiding Sweetgrass: Indigenous Wisdom, Scientific Knowledge, and the Teaching of Plants*. Minneapolis, MN: Milkweed Editions.

Kirksey, Eben, Nicholas Shapiro, and Maria Brodine. 2014. "Hope in Blasted Landscapes." Pp. 29–63 in *The Multispecies Salon*, edited by Eben Kirksey. Durham, NC: Duke University Press.

Kniffen, Fred B., Hiram F. Gregory, and George A. Stokes. 1987. *The Historic Indian Tribes of Louisiana from 1542 to the Present*. Baton Rouge, LA: Louisiana State University Press.

Kniffen, Fred Bowerman, and Samuel Hilliard. 1988. *Louisiana: Its Land and Its People*. Baton Rouge, LA: Louisiana State University Press.

Knight, William. 2016. "Blurring the Boundaries: Subsistence and Recreational Fisheries in Late-Nineteenth Century Ontario." Pp. 60–75 in *Subsistence under Capitalism: Historical and Contemporary Perspectives*, edited by James Murton, Dean Bavington, and Carly Dokis. Montreal, CA: McGill-Queen's University Press.

Knight, William. 2009. "Settler Subsistence Fisheries in 19th Century Ontario." Paper presented at the NICHE workshop "Bringing Subsistence out of the Shadows," North Bay Ontario, October 2009, http://nich-canada.org/resources/conference-workshop-archive/bringing-subsistence-out-of-the-shadows.

Kopytoff, Igor. 1986. "The Cultural Biography of Things: Commoditization as Process." Pp. 64–91 in *The Social Life of Things: Commodities in Cultural Perspective*, edited by Arjun Appadurai. Cambridge, UK: Cambridge University Press.

Kuokkanen, Rauna. 2011. "Indigenous Economies, Theories of Subsistence, and Women: Exploring the Social Economy Model for Indigenous Governance." *American Indian Quarterly* 35(2):215–40.

Le Page du Pratz, Antoine-Simon. 1947 (1774). *The History of Louisiana, or of the Western Parts of Virginia and Carolina: Containing a Description of the Countries that Lie on Both Sides of the River Mississippi: With an Account of the Settlements, Inhabitants, Soil, Climate, and Products*. New Orleans, LA: J. S. W. Harmanson.

Lee, Richard. 2007. "Food Security and Food Sovereignty." Centre for Rural Economy Discussion Paper Series No. 11. Center for Rural Economy, University of Newcastle upon Tyne, pp. 1–17.

Li, Judith, ed. 2007. *To Harvest, to Hunt: Stories of Resource Use in the American West*. Corvallis, OR: Oregon State University Press.

Louisiana Department of Health and Human Resources. 2014a. Region 3 Parish Community Health Assessment Profile: Lafourche Parish. Available at: http://lphi.org/CMSuploads/Lafourche-Parish_Region-3-Parish-Community-Health-Assessment-Profile-60366.pdf.

Louisiana Department of Health and Human Resources. 2014b. Region 3 Parish Community Health Assessment Profile: Terrebonne Parish. Available at: http://lphi.org/CMSuploads/Terrebonne-Parish_Region-3-Parish-Community-Health-Profile-60470.pdf.

Maciage, Michael. 2019. "Born and Raised: The Parts of America with the Most Natives." *Governing.com*. Retrieved April 20, 2022 from https://www.governing.com/archive/native-homegrown-residents-by-county.html.

MacKinnon, Danny, and Kate Driscoll Derickson. 2013. "From Resilience to Resourcefulness: A Critique of Resilience Policy and Activism." *Progress in Human Geography* 37(2):253–70.

MacRae, Graeme. 2016. "Food Sovereignty and the Anthropology of Food: Ethnographic Approaches to Policy and Practice." *Anthropological Forum* 26(3):227–32.

Maldonado, Julie K., Christine Shearer, Robin Bronen, Kristina Person, and Heather Laszrus. 2013. "The Impact of Climate Change on Tribal Communities in the US: Displacement, Relocation, and Human Rights." *Climate Change* 120:601–14.

Marks, Brian. 2012. "The Political Economy of Household Commodity Production in the Louisiana Shrimp Fishery." *Journal of Agrarian Change* 12(2 & 3):227–51.

Marshall, Bob, The Lens, Brian Jacobs, and Al Shaw. 2014. "Losing Ground." *ProPublica*, August 28, 2014. Available at: https:/project.propublica.org/louisiana/.

Masson, Todd. 2017. "Two Men Were 42 Redfish over Limit, Agents Say." *Times-Picayune/NOLA.com*, January 12, 2017.

Mathews, Gordon, and Carolina Izquierdo. 2009a. "Introduction: Anthropology, Happiness, and Well-Being." Pp. 1–9 in *Pursuits of Happiness: Well-being in Anthropological Perspective*. New York, NY: Berghahn Books.

Mathews, Gordon, and Carolina Izquierdo. 2009b. "Conclusion: Towards an Anthropology of Well-Being." Pp. 248–66 in *Pursuits of Happiness: Well-being in Anthropological Perspective*. New York, NY: Berghahn Books.

Matsutake Worlds Research Group. 2009. "A New Form of Collaboration in Cultural Anthropology." *American Ethnologist* 36(2):380–403.

Mattingly, Cheryl. 2014. *Moral Laboratories: Family Peril and the Struggle for a Good Life*. Berkeley, CA: University of California Press.

McCaffrey, Kevin, and Neil Alexander. 2011. *We Live to Eat: New Orleans Love Affair with Food*. New Orleans, LA: ePrime Media.

McGuire, Tom. 2008. *History of the Offshore Oil and Gas Industry in Southern Louisiana*. Volume II: *Bayou Lafourche—Oral Histories of the Oil and Gas Industry*. US Department of the Interior, Minerals Management Service, Gulf of Mexico OCS Region, New Orleans, LA. OCS Study MMS 2008-043.

McGuire, Tom, and Diane Austin. 2013. "Beyond the Horizon: Oil and Gas along the Gulf of Mexico." Pp. 298–311 in *Cultures of Energy: Power, Practices, Technologies*, edited by Sarah Strauss, Stephanie Rupp, and Thomas Love. Chicago, IL: Left Coast Press.

McLain, Rebecca J., Peter Hurley, Marla R. Emery, and Melissa R. Poe. 2014. "Gathering 'Wild' Food in the City: Rethinking the Role of Foraging in Urban Ecosystem Planning and Management." *Local Environment* 19(2):220–40.

McTighe, Laura, and Megan Raschig. 2019. "Introduction: An Otherwise Anthropology." Theorizing the Contemporary. *Fieldsights*, July 31. https://culanth.org/fieldsights/introduction-an-otherwise-anthropology.

Mehta, Jayur Madhusudan, and Elizabeth Chamberlain. 2019. "Mound Construction and Site Selection in the Lafourche Subdelta of the Mississippi River Delta, Louisiana, USA." *Journal of Island and Coastal Archaeology* 14(4):453–78. DOI: 10.1080/15564894.2018.1458764.

Menzies, Charles R. 2010. "Dm Sibilhaa'nm da Laxyuubm Gitxaala: Picking Abalone in Gitxaala Country." *Human Organization* 69(3):213–20.

Mihesuah, Devon A., and Elizabeth Hoover, ed. 2019. *Indigenous Food Sovereignty in the United States: Restoring Cultural Knowledge, Protecting Environments, Regaining Health*. Norman, OK: University of Oklahoma Press.

Mol, Annemarie, Ingunn Moser, and Jeannette Pols, eds. 2010. *Care in Practice: On Tinkering in Clinics, Homes, and Farms*. Bielefeld: Transcript-Verlag (MatteRealities / VerKorperungen: Perspectives from Empirical Science Studies).

MQVN. 2010. Loss of Subsistence Use Claim Framework and Template for Louisiana Vietnamese American Fisherfolk & Other Louisiana Fisherfolks. Impact Claim

Project White Paper. Mary Queen of Vietnam Community Development Corporation: New Orleans.

Murton, James, Dean Bavington, and Carly Dokis, eds. 2016. "Introduction: Why Subsistence?" Pp. 3–39 in *Subsistence under Capitalism: Historical and Contemporary Perspectives*. Montreal, CA: McGill-Queen's University Press.

Newhouse, David R. 2000. "Resistance Is Futile: Aboriginal Peoples Meet the Borg of Capitalism." Pp. 141–55 in *Ethics and Capitalism*, edited by John Douglas Bishop. Toronto, CA: University of Toronto Press.

Owens, Maida, and Eileen Engel. 2011. "A Tale of Discovery: Folklorists and Educators Collaborate to Create and Implement the Louisiana Voices Educators Guide." Pp. 47–67 in *Through the Schoolhouse Door: Folklore, Community, Curriculum*, edited by Paddy Bowman and Lynne Hamer. Logan, UT: Utah State University Press.

Ownby, Ted. 1990. *Subduing Satan: Religion, Recreation, and Manhood in the Rural South, 1865–1920*. Chapel Hill, NC: University of North Carolina Press.

Oyunginka, Ebenezer O., David R. Lavergne, and Latika Bharadwaj. 2011. Louisiana Commercial Shrimp Fishermen: Trends in Fishing Efforts, Landings and Landing Revenue, Impact of Hurricanes and Monitoring of Recovery. Louisiana Department of Wildlife and Fisheries, Office of Fisheries Socioeconomic Research and Development Section. Baton Rouge, Louisiana. Available at: http://www.wlf.louisiana.gov/sites/default/files/pdf/page/37774-economic-reports/tripticketshrimp-report.pdf.

Parfait, Jessi. 2019. The Effects of Forced Migration on Houma Culture. MA thesis, Department of Geography and Anthropology, Louisiana State University, Baton Rouge.

Pew Research Center. 2009. "Magnet or Sticky?: A State-by-State Typology." Available at: http://www.pewsocialtrends.org/2009/03/11/magnet-or-sticky/.

Pierre-Louis, Kendra. 2021. "The Fossil Fuel's Legacy of White Supremacy: The Links Extend beyond the Corporate Office and Drill Pad." *Sierra*, April 2, 2021. https://www.sierraclub.org/sierra/fossil-fuel-industrys-legacy-white-supremacy.

Poe, Melissa R., Phillip S. Levin, Nick Tolimieri, and Karma Norman. 2015. "Subsistence Fishing in a 21st-Century Capitalist Society: From Commodity to Gift." *Ecological Economics* 116:241–50.

Poe, Melissa R., Rebecca J. McLain, Marla R. Emery, and Peter T. Hurley. 2013. "Urban Forest Justice and the Right to Wild Foods, Medicines, and Materials in the City." *Human Ecology* 41:409–22.

Pottery, Nancy. 2016. "Fishing for Subsistence, Sport, and Sovereignty on Lake Nipissing." Pp. 147–61 in *Subsistence under Capitalism: Historical and Contemporary Perspectives*, edited by James Murton, Dean Bavington, and Carly Dokis. Montreal, CA: McGill-Queen's University Press.

Prentice, Rebecca. 2012. "No One Ever Showed Me Nothing: Skill and Self-Making among Trinidadian Garment Workers." *Anthropology and Education Quarterly* 43(4):400–414.

Price, Richard, ed. 1996 [1973]. *Maroon Societies: Rebel Slave Communities in the Americas*. Baltimore, MD: Johns Hopkins University Press.

Priest, Tyler, and Michael Botson. 2012. "Bucking the Odds: Organized Labor in Gulf Coast Oil Refining." *Journal of American History* 99(1):100–110.

Probyn, Elspeth. 2016. *Eating the Ocean*. Durham, NC: Duke University Press.

Puig de la Bellacasa, Maria. 2015. "Making Time for Soil: Technoscientific Futurity and the Pace of Care." *Social Studies of Science* 45(5):691–716.

Putnam, Robert. 1990. *Bowling Alone: The Collapse and Revival of American Community*. New York, NY: Simon & Schuster.

Rappaport, Roy. 1993. "Distinguished Lecture in General Anthropology: The Anthropology of Trouble." *American Anthropologist* 92(2):295–303.

Rawal, Sanjay. 2020. *Gather*. Produced by Tanya Ager Meillier and Sterlin Harjo. Boulder, CO: First Nations Development Institute/Ager Meillier Films/Illumine.

Reese, Ashanté M. 2019. *Black Food Geographies: Race, Self-Reliance, and Food Access in Washington, DC*. Chapel Hill, NC: University of North Carolina Press.

Reese, Ashanté M. 2018. "'We will not perish; we're going to keep flourishing': Race, Food Access, and Geographies of Self-Reliance." *Antipode* 50(2):407–24.

Regis, Helen, and Shana Walton. 2019. "From Festivals to Subsistence and Back Again." Paper presented in *Re-Thinking Subsistence in Turbulent Times: New Contexts, Configurations, and Intersections with Social and Environmental Justice*, organized by Shirley Fiske and Patricia Clay, Annual Meeting of the Society for Applied Anthropology, Portland, Oregon, March 22, 2019.

Regis, Helen A., and Shana Walton. 2015. *Subsistence in Coastal Louisiana, Volume I: An Exploratory Study*. US Department of the Interior, Bureau of Ocean Energy Management, Gulf of Mexico OCS Region, New Orleans, LA. OCS Study BOEM 2015.

Robbins, Paul, Marla Emery, and Jennifer L. Rice. 2008. "Gathering in Thoreau's Back Yard: Nontimber Forest Product Harvesting as Practice." *Area* 40(2):265–77.

Roberts, Faimon. 2021. "After Hurricane Ida destroyed 1 in 4 Grand Isle buildings, demolition is on 'path to recovery.'" nola.com. December 1, 2021. Accessed April 6, 2022: https://www.nola.com/news/environment/article_921438d2-52cc-11ec-9fa7-1ff714e37988.html.

Roman-Alcalá, Antonio. 2013 "From Food Security to Food Sovereignty." *Civil Eats*. Digital resource. Accessed March 21, 2019.

Salmón, Enrique. 2012. *Eating the Landscape: American Indian Stories of Food, Identity, and Resilience*. Tucson, AZ: University of Arizona Press.

Samuel, Sajay. 2016. "In Defense of Vernacular Ways." Pp. 318–45 in *Subsistence under Capitalism: Historical and Contemporary Perspectives*, edited by James Murton, Dean Bavington, and Carly Dokis. Montreal, CA: McGill-Queen's University Press.

Sayers, Daniel O. 2016. *A Desolate Place for a Defiant People: The Archaeology of Maroons, Indigenous Americans, and Enslaved Laborers in the Great Dismal Swamp*. Gainesville, FL: University Press of Florida.

Schemerhorn, Clavin. 2017. "The Thibodaux Massacre Left 60 African-Americans Dead, and Spelled the End of Unionized Farm Labor in the South for Decades." *Smithsonian Magazine*. November 21, 2017. Digital edition, accessed March 3, 2022: https://www.smithsonianmag.com/history/thibodaux-massacre-left-60-african-americans-dead-and-spelled-end-unionized-farm-labor-south-decades-180967289/.

Schleifstein, Mark. 2020. "'We're Screwed': The Only Question Is How Quickly Louisiana Wetlands Will Vanish, Study Says." Nola.com. May 22.

Schuller, Mark. 2012. *Killing with Kindness: Haiti, International Aid, and NGOs*. New Brunswick, NJ: Rutgers University Press.

Schuller, Mark. 2016. *Humanitarian Aftershocks in Haiti*. New Brunswick, NJ: Rutgers University Press.

Scott, Rebecca. 2005. *Degrees of Freedom: Louisiana and Cuba after Slavery*. Cambridge, MA: Harvard University Press.

Sell, James L., and Tom McGuire. 2008. *History of the Offshore Oil and Gas Industry in Southern Louisiana. Volume IV: Terrebonne Parish*. US Department of the Interior,

Minerals Management Service, Gulf of Mexico OCS Region, New Orleans, LA. OCS Study MMS 2008–045.

Singer, Merrill, and Barbara Rylko-Bauer. 2021. "The Syndemics and Structural Violence of the COVID Pandemic: Anthropological Insights on a Crisis." *Open Anthropological Research* 1:7–32.

Solet, Kimberly. 2006. Thirty Years of Change: How Subdivisions on Stilts Have Altered a Southeast Louisiana Parish's Coast Landscape and People. Master's thesis, Department of Planning and Urban Studies, University of New Orleans, New Orleans, Louisiana.

Spitzer, Nick. 1985. *Louisiana Folklife (A Guide to the State)*. Published by the Louisiana Folklife Program, Division of the Arts, Department of Culture, Recreation and Tourism.

St. John, Graham. 2012. "Altered Together, Dance Festivals, and Cultural Life. Vital Topics Forum: Happiness." *American Anthropologist* 114(1):9–10.

Syvitski, James P. M., Albert J. Kettner, Irina Overeem, Eric W. H. Hutton, Mark T. Hannon, G. Robert Brakenridge, John Day, Charles Vörösmarty, Yoshiki Saito, Liviu Giosan, and Robert J. Nicholls. 2009. "Sinking Deltas Due to Human Activities." *Nature Geoscience* 2:681–86. DOI: 10.1038/ngeo629.

Teitelbaum, Sara, and Thomas M. Beckley. 2006. "Harvested, Hunted, and Home Grown: The Prevalence of Self-Provisioning in Rural Canada." *Journal of Rural and Community Development* 1:114–30.

Thin, Neil. 2008. "Realizing the Substance of Their Happiness: How Anthropology Forgot about *Homo Gauiusus*." Pp. 134–55 in *Culture and Well-Being: Anthropological Approaches to Freedom and Political Ethics*, edited by Alberto Corsín Jiménez. London, UK: Pluto.

Tidwell, Mike. 2004. *Bayou Farewell: The Rich Life and Tragic Death of Louisiana's Cajun Coast*. New York, NY: Vintage.

Trépanier, Cécyle. 1991. "The Cajunization of French Louisiana: Forging a Regional Identity." *Geographical Journal* 157(2):161–71.

Tsing, Anna for the Matsutake Worlds Research Group. 2016. "Blasted Landscapes and the Gentle Arts of Mushroom Picking." Pp. 87–111 in *The Multispecies Salon*, edited by Eben Kirksey. Durham, NC: Duke University Press.

Tsing, Anna Lowenhaupt. 2015. *The Mushroom at the End of the World: On the Possibility of Life in Capitalist Ruins*. Durham, NC: Duke University Press.

Ulysse, Gina A. 2007. *Downtown Ladies: Informal Commercial Importers, A Haitian Anthropologist, and Self-Making in Jamaica*. Chicago, IL: University of Chicago Press.

Underhill, Ruth. 1938. Report on a Visit to Indian Groups in Louisiana, October 15–25, 1938. US Bureau of Indian Affairs.

United Houma Nation. n.d. History. Available at: unitedhoumanation.org.

US Census Bureau. 1850. Census of Population and Housing 1850. Available at: https://www.census.gov/prod/www/decennial.html.

US Census Bureau. 2010. Census Demographic Profiles for Louisiana, Lafourche, and Terrebonne Parishes. Available at: http://2010.census.gov/2010census/popmap/.

US Census Bureau. 2023. Growth in Nation's Largest Counties Rebounds in 2022. Available at: https://www.census.gov/newsroom/press-releases/2023/population-estimates-counties.html.

US Department of Agriculture. 2012. Census of Agriculture. Parish Profile. Lafourche Parish. Available at: http://www.agcensus.usda.gov/Publications/2012/Online_Resources/County_Profiles/Louisiana/cp22057.pdf.

US Department of Agriculture. 2012. Census of Agriculture. Parish Profile. Terrebonne Parish. Available at: http://www.agcensus.usda.gov/Publications/2012/Online_Resources/County_Profiles/Louisiana/cp22109.pdf.

Usner, Daniel H., Jr. 2018. *American Indians in Early New Orleans: From Calumet to Raquette*. Baton Rouge, LA: Louisiana State University Press.

Usner, Daniel H., Jr. 1986. "Food Marketing and Interethnic Exchange in the 18th Century Lower Mississippi Valley." *Food & Foodways* 1:279–310.

Verdin, Monique. 2020. *Botanica Series*. New Orleans, LA: The Land Memory Bank and Seed Exchange and the Neighborhood Story Project.

Verdin, Monique. 2019. *Return to Yakni Chitto: Houma Migrations*, photographs and essays by Monique Verdin, in conversation with T. Mayheart Dardar, Anesie and Jane Verdin, Allison Rodriguez, Raymond Moose Jackson, Kathy Randels, and Nick Slye. Edited by Rachel Breunlin. New Orleans, LA: UNO Press / Neighborhood Story Project.

Verdin, Monique. 2013. "Ebb and Flow: Migrations of the Houma, Erosions of the Coast / Southward into Vanishing Lands." Pp. 19–24 in *Unfathomable City: A New Orleans Atlas*, edited by Rebecca Solnit and Rebecca Snedeker. Berkeley, CA: University of California Press.

Verdin, Monique, and Rachel Breunlin. 2020. On Remembering: Conversation and Poetry in Honor of Indigenous People's Day. Featuring Monique Verdin, Rachel Breunlin, and Raymond "Moose" Jackson. Video. New Orleans: Newcomb Art Museum, October 12, 2021. Digital resource, accessed August 1, 2021. https://newcombartmuseum.tulane.edu/portfolio-item/on-remembering/.

Verdin, Monique, and Sharon Linezo Hong, dirs. 2013. *My Louisiana Love*. Produced by Sharon Linezo Hong, Julie Mallozzi, and Monique Michelle Verdin. Written by Monique Verdin and Sharon Linezo Hong. Berkeley, CA: Berkeley Media.

Wagner, Bryan. 2019. *The Life and Legend of Bras-Coupé: The Fugitive Slave Who Fought the Law, Ruled the Swamp, Danced at Congo Square, Invented Jazz and Died for Love*. Baton Rouge, LA: Louisiana State University Press.

Wali, Alaka. 2012. "A Different Measure of Well-Being." *American Anthropologist* 114(1):16–18.

Wali, Alaka, Diana Alvira, Paula S. Tallman, Ashwin Ravikumar, and Miguel O. Macedo. 2017. "A New Approach to Conservation: Using Community Empowerment for Sustainable Well-Being." *Ecology and Society* 22(4):6.

Walker, Harry, and Iza Kavedzija. 2015. "Values of Happiness." *Hau: Journal of Ethnographic Theory* 5(3):1–23.

Walton, Shana. 2003. "Louisiana's Coonasses: Choosing Race and Class over Ethnicity." Pp. 38–50 in *Signifying Serpents and Mardi Gras Runners: Representing Identity in Selected Souths*, edited by Celeste Ray and Luke Eric Lassiter. Southern Anthropological Society Proceedings No. 36. Athens, GA: University of Georgia Press.

Walton, Shana. 2002. "Not with a Southern Accent: Cajun English and Ethnic Identity." Pp. 104–19 in *Linguistic Diversity in the South: Changing Codes, Practices, and Ideology*, edited by Margaret Clelland Bender. Athens, GA: University of Georgia Press.

Walton, Shana. 1994. Flat Speech and Cajun Ethnic Identity in Terrebonne Parish, Louisiana. PhD dissertation, Department of Anthropology, Tulane University, New Orleans, Louisiana.

Walton, Shana, and Helen A. Regis. 2015. "This Is What We Do Here: Subsistence Practices in Coastal Louisiana." Keynote Lecture, Louisiana Academy of Sciences. Thibodeaux, LA, March 13, 2015.

Westman, Clinton N. 2016. "Aboriginal Subsistence Practices in an 'Isolated' Region of Northern Alberta." Pp. 162–94 in *Subsistence under Capitalism: Historical and Contemporary Perspectives*, edited by James Murton, Dean Bavington, and Carly Dokis. Montreal, CA: McGill-Queen's University Press.

Wilmore, Cherry. 2022. Panelist for "Community Testimonials from the Bayou Region." Managed Retreat in the U.S. Gulf Coast Region, a workshop by the National Academy of Sciences, Engineering, and Medicine. July 28, 2022.

Wurzlow, Helen. 1984. *I Dug Up Houma, Terrebonne*. Houma, LA. Self-published.

Yim, Paige. 2019. Crabs = Happiness. Blog entry. Digital resource. http://caughtncooked.com/post/31202779184/crabs-happiness-its-true-the-documentary.

Young, Amy L., Michael Tuma, and Cliff Jenkins. 2001. "The Role of Hunting to Cope with Risk at Saragossa Plantation, Natchez." *American Anthropologist* 103(3):692–704.

INDEX

abundance, 11, 109, 137, 157, 158, 179n9
Acadians, 36, 42, 43, 45, 49, 55, 59, 177n5. *See also* Cajuns
Adams, Chris, 16, 27, 121, 123, 165
African Americans, 7; crabbers, 141; fishers, 58; heritage, 53; history, 56–60; identity, 59–60, 178n8; labor organizing, 57; occupations, 50, 56; population, 38; self-provisioning, 44; subsistence practices, 44, 60, 168, 179n3. *See also* maroons
African people, 36, 41, 43, 52
agency, 148
agriculture, 4, 36, 37, 39, 43, 44, 56, 157; cash crops, 42, 44, 50
alligators, 9, 11, 15, 19, 21–23, 39, 51–53, 68, 87, 96, 118, 139, 150, 159, 179n6
American Indians, 7, 40, 47, 49, 53–55, 60, 177n4, 178n3. *See also* Indigenous people
Anglo-Americans, 36
Angola (Louisiana State Penitentiary), 40
archaeology, 38–41, 44
Asian people, 9, 178n7 (chap. 3)
asparagus, 27–30, 112, 114
assemblages, 19, 148, 159, 182n5
Atchafalaya River, xi, 33, 36, 47, 65, 179n2
Austin, Diane, 12, 165
Autin, Joe and Claudia, 68–69
autonomy, 9, 16, 117, 145, 148, 159

BARA (Bureau of Applied Research in Anthropology), 12
barter, x, 4, 10, 96, 120–21

Bayou Farewell, 56
Bayougoula (tribe), 40
bayous, 7, 33, 39, 51, 54, 60; Bayou Culture Collaborative, 18, 164, 178n4, 198; Bayou Dularge, 54; Bayou Grand Caillou, 19, 39; Bayou Lafourche, 39–42, 44–45; Bayou L'Ourse, 70; Bayou Pointe-aux-Chênes, xii; Bayou Sherman, 79–82; Bayou St. John, ix; Bayou Teche, 40; Bayou Terrebonne, 43; Bayou Woman, 16; Center for Bayou Studies, 12, 16, 165, 198; etymology, 39; Little Caillou, 54
bees, 27, 30
belonging, sense of, 17, 22, 77, 104, 157
Bennett, Effie, 60
Bergeron, Arthur, 70, 71, 99–101, 130, 149
Bienville, Louisiana, 40, 41
Billiot, Alexandre, 43
Billiot, Jean, 42
Billiot, Wendy Wilson, vii–xi, 16, 60, 93, 107–8, 113, 117, 118, 143, 148, 164, 165
birdfoot delta, 7, 33, 40
boat blessings, 17, 19, 21, 140, 176n1
Bobtown, Louisiana, 58
BOEM (Bureau of Ocean Energy Management and Research), vii, 12, 17, 86, 162, 165
Borne, Richard, 70, 72–74, 79–86, 90, 95, 125–26, 152, 166
boucheries, 19, 21–22, 30, 176n2
Bourdieu, Pierre, 104
Bourg, Claude, 55
Bourg, Jenny, 129

Index

Bourg, Louisiana, 67, 123, 178n2
Browne, Katherine, 115
Brule Hunting Club, 79–83

Cajuns, 55–56, 59; Cajunization, 56. *See also* Acadians
Callaway, Don, 6, 58, 59, 157, 165, 166, 173, 179n3
camps, 6, 15, 25–27, 30, 46, 48, 67, 69–72, 74, 75, 79–83, 87, 90–97, 102, 104, 108, 119, 125–27, 130, 139, 148, 150, 157, 160, 164, 168, 169, 176n5 (chap. 2)
canals, xii, 7; catching crabs, 66; and coastal erosion, 175n4; dug by lumber companies, 50; and limits to fishing access, 85, 87; by oil and gas, 50
capitalism, 3, 101, 132; global, 7, 170
care, 101, 115, 132, 135, 142–45, 154, 157–59, 162, 164
cash: below-market exchange, 120; crops, 4, 42, 44, 50; economy, 48, 52; for corn and tomatoes, 119; for shrimping license, 9; roadside stands, 121; sale of seafood, 157
Celestin, Bob, 58
Center for Bayou Studies, 12, 16, 165, 198
Chacahoula, Louisiana, 99
Chaisson, Lora Ann, 16, 51, 52, 91, 163, 165
Chauvin, Louisiana, 17, 19, 20, 21, 33, 58, 178n2; folk festival, 65, 153
Choctaw language, 39, 135
citrus, 50, 109, 110–14, 122, 123, 175, 181
Clay, Patricia, 156
climate change, 5, 6, 161, 170
clubs, 13, 71, 79–87, 89–90, 95–97, 102, 119, 139, 152, 169
coastal erosion, 175n4. *See also* land loss
colonization, 41–42
Community Economies Collective, 7
community economy, 157
conservation, 46, 85, 151–52, 156
corn, 11, 39, 41, 43, 44, 52, 54, 99–101, 119–20, 125, 179n9
Cory (student), 75, 76, 130, 150
Courteau, Rosalie, 33
crabs, 19, 24–27, 37, 53, 56, 104–9, 114, 116–17, 122, 154, 162, 168, 179n1; crabbing, 11, 15, 68, 72, 73–75, 120, 125, 128, 141, 143, 148, 158; "Crabs Under a Microscope," 63–66, 159
crawfish, 21, 22, 28, 65, 113
culture: African American, 60; Cajun, 55–56; enculturation, 27, 65; family gatherings, 67; folk, 171; French-speaking, 45; harvesting, 76; hunting, 79; land, plants, and culture, 54; living off the land, 64; and nature, 7; nature-cultures, 175n4; North American, viii; people of other cultures, 117; plantation, 50; popular, 36, 83, 177n7; and resilience, 178n2; and seasonal migration, 49

Dawdy, Shannon, 4, 41
Deepwater Horizon disaster, 12, 147, 160
deer (marsh deer), 11, 15, 20, 21, 25, 26, 39, 67, 68–70, 75, 79–90, 95–97, 105–9, 113, 118, 125–28, 130–32, 144, 151, 160, 171, 180nn3–4
DeHart, Esmiel, 48
Digilormo, Jamie, 17, 131, 141, 165
disasters, 12, 154, 160, 163, 182n2
diseases, 38, 115, 177n2; blue tongue, 81, 85, 180n4 (chap. 6); congestive heart failure, 37; decimating Indigenous populations, 41; diabetes, xi, 5, 38, 175n1; links to poverty, 37–38. *See also* health
displacement, 7, 33, 36, 41, 45
distributary, xiii, 7, 38, 39
diverse economies, 9, 101
Donaldsonville, Louisiana, 42
duck hunting, x, 11, 15, 20–21, 25, 39, 46–48, 61, 69–70, 75, 87, 89, 92, 95, 97, 105–7, 111, 118, 120, 125–28, 131–33, 137, 143, 144, 148, 151, 152, 162, 179n8, 180n3, 181n2
Ducks Unlimited, 152
Duet, Tiffany, 17, 18, 22, 92, 99, 101, 103, 124, 165
Dulac, Louisiana, 24, 43, 51, 53, 58, 91, 92, 115, 178n2
Duncan, Colin, 132
Dupré, Mrs., vii, ix–xi, 3, 5, 68, 113, 114, 128, 143, 157

elderberries, 135
enslaved Africans, 36, 41, 43, 44, 52, 57, 58, 168, 170
environmental justice, 166
environmental knowledge, 5, 85, 151, 158, 161, 165
environmental stewardship, 139, 151–52
Eschete, Rory, 25–27, 64, 67, 131, 150, 168
ethnicity: African American, 7, 59–60, 178n8; Cajun, 55–56, 59; European American, 7, 38, 53, 55, 56, 57, 59, 87; Filipino, 52, 53, 176n3, 177n7; German, 7, 36, 41, 45; Irish, 7, 45; Isleño, 7, 177n5; Italian, 28, 45, 52, 57; Latin American, 36, 38, 53; Laotian, 7, 36, 53; Sicilian, 7, 45; Vietnamese, 7, 36, 53, 105. *See also* race
ethnography, 65, 153; team ethnography, 12–13, 16–18. *See also* research methods
exchange, 119–24

fast food, 24, 170
feasts, 63–78; boucheries, 19, 21–22, 30, 176n2; community dinners, 21, 71, 108; dinners, 141. *See also* crabs
Felicia (interviewee), 92–93, 96, 102–4, 119, 148
festivals, viii, 13, 15, 17, 51, 171; Blessing of the Fleet (boat blessing), 17, 19, 21, 140, 176n1
filé, 52, 53, 122, 124, 136, 159, 181. *See also* sassafras
First Nations, 4. *See also* Indigenous people
fishing, ix, 4, 6, 26, 28, 37, 46–48, 53, 54, 59, 60–61, 65, 67, 69, 71–75, 83, 85, 87, 89, 92, 95, 97, 101, 102–5, 107, 113, 118, 125, 127, 131, 136, 138, 141, 151, 155–57, 160, 162, 163, 170, 179n9, 180n6; commercial, 10, 61, 97, 101, 105, 107–8, 151, 155–57, 179; fried fish, 21, 25, 70, 113, 114, 117; regulations, 155; sports fishing, 107, 108, 155, 161. *See also* shrimping
Fisk, Harold, 8–9
Fiske, Shirley, 6, 58, 59, 156–57, 165, 166, 173, 179n3
flooding, 39, 94, 129, 161, 180n5 (chap. 7)
food: deserts, 115; justice, 154; logs, 108–17; security, 154; sovereignty, 5, 9, 11, 153–55, 159, 176n12, 181n1 (chap. 9)

Foret, Jonathan, 164, 165, 182n3
freedmen, 46
French people: colonization, 41–42; cultural assimilation, 36; descent, 7; French fries, 117; language, 45, 51, 60, 176n2, 177n8; Louisiana, 122; names, 56, 60; nationality, 28; onion soup, 52; pronunciation, 67; rabbit recipe, 30; settlers, 49; speakers, 55, 59. *See also* Acadians
French Quarter (New Orleans, Louisiana), 40
frog gigging, 25, 26, 61, 82, 107
freezers, 11, 13, 15, 16, 20, 22, 67, 69, 89, 90, 97, 100, 104, 105, 106, 107, 108, 113, 120, 125, 128, 144, 148, 155, 171
fruit, 11, 108, 109, 111, 113, 119, 135

Galeucia, Annemarie, ix, 17, 23, 68, 165
gallicization, 45
gardening: contribution to meals, 113; diversity, 114; ecological approach, 28; as heritage, 162, 168, 171; interplanting, 30; investment, 100; learned, 70; linked to identity, 53, 61; network of indigenous gardens, 135; as part of a constellation of subsistence practices, 15, 19, 22; as part of a mixed economy, 106, 120; as reclaiming connections, 135; as reclaiming land, 144; renewed interest in, 5; significance, 9, 161; talking about, 137; and well-being, 156
Gayarre, Charles, 41
gender identity, 9, 10, 150, 169
German people, 7, 36, 41, 45
Gibson-Graham, J. K., 7, 101, 176n7
gifts, xi, 4, 10, 19, 30, 104, 106, 108, 120, 125, 162
global connections, 3–4, 10, 18, 30, 176n3
global commodity, 18
global economies, intersections with local foodways, 6
global markets, 60
global warming, 161
Golden Meadow, Louisiana, 33, 135
Grand Isle, Louisiana, 26, 27, 43, 49, 51, 53, 54, 55, 67, 71, 72, 91, 92, 96, 168, 176n6

Greer, Tammy, 54, 166, 178n4
Griffin, Mary Ann, 143
Guarisco, Al, 27–31, 70

Hall, Gwendolyn Midlo, 44
happiness, 155–59. *See also* well-being
harvesting, viii, 5, 7, 19, 24, 56, 65, 68, 72, 76, 83, 137, 153, 171, 179n1; care, 157; cost, 106, 132; history, 33–50; identity, 140–42, 150–51; importance, 162; logs, 108–15, 122, 130; making people, 136; mutual aid, 161; overharvesting, 152; practice, 180n1. *See also* conservation
health: and access to clean water, 138; challenges, 115; codes, 20; focus on individual, 145; and food sovereignty, 5, 181n1; and healthy foods, 115–16, 133; impact of environmental crises, 160; indicators, 37–38; mental, 27; social and biological dimensions, 177n2; structural violence, 177n2
Henry, Eddie, 46
Hingle, Paul, 86
history, 4, 9, 12, 21, 33–50, 53–54, 57–61, 65, 86, 139, 168, 176n12, 179n6. *See also* oral history
Hochschild, Arlie Russell, 161
Houma (United Houma Nation), 16, 17, 33, 36, 37, 49, 50, 51, 53, 54, 114, 135, 145, 161, 181n1 (chap. 8). *See also* Indigenous people
Houma, Louisiana, 28, 40
Hubbard, Audriana, 17, 19–22, 24, 139, 140–41, 148, 150, 165, 176n1
Hunters for the Hungry, 143
hunting: by freedmen, 46; as recreation, 46; regulations, 15, 16, 46, 136, 168, 171, 179n8; as sport, 46, 87, 143, 161; as tourism, 46, 158. *See also* camps
hurricanes, 94, 95, 159, 160, 161, 163; Hurricane Ida, 163–64

identity, 5, 9, 11, 19, 22, 27, 36, 37, 51–61, 76, 83, 104, 105, 140–42, 145, 150, 154, 157, 159, 161, 168, 169, 178n8
Indigenous people, 4–7, 17, 18, 33–37, 38–41, 44, 51–55, 59, 135–36, 155–56, 159, 168, 175n2 (chap. 2), 177n4, 178n3, 181n1; Acolapissa, 42; Alaska Natives, 5, 6, 73, 153, 173; Atakapa, 41; Bayougoula, 40; Biloxi, 40–42; Biloxi-Chitimacha Confederation of Muskogee, 43; Chawasha, 40, 41; Chitimacha, 40, 41, 42, 43, 53; Grand Caillou-Dulac Band, 49, 53; Houma (United Houma Nation), 16, 17, 37, 49, 50, 51, 53, 54, 114, 135, 145, 161, 181n1 (chap. 8); Isle de Jean Charles Band, 49, 53; Natchez, 41; Nipissing Nation, 155; Point-au-Chien Indian Tribe, 49, 53, 148, 175n3; Seminole, 36; Tunica, 41, 42; Washa, 40–41; Yakama Nation, 5
inequality, 38, 116, 148, 177n2. *See also* poverty
informal economy, 3, 143
investments, 125–30
Isle de Jean Charles, Louisiana, 43, 51, 54, 67, 115, 144, 163
Isle de Jean Charles Band, 49, 53
Isleño people, 7, 177n5
isolation, 7, 9, 22, 30, 39, 91; stereotype, 30–31, 83. *See also* global connections

Jacob (student), 63–67, 159, 179n5
Jacob, Michelle, 5
jambalaya, 21, 51–52, 56, 114, 121, 139, 178n1

Kimmerer, Robin Wall, 5
Klondyke, Louisiana, 50
Knight, William, 155
Knights of Labor, 57
Kuokkanen, Rauna, 138

Labadieville, Louisiana, 42, 79, 80, 180n1
Lafourche Parish, ix, 11, 12, 16, 17, 24, 26, 28, 30, 31, 33, 37–45, 48–52, 57–58, 61, 68–70, 87, 89, 97, 99, 121–23, 132, 136, 152, 158, 161, 164, 175n3, 177n5, 178n8
lagniappe, 58, 178n6
land loss, 6, 61, 87, 88, 89, 97, 152, 159–61, 164, 175n4
Land Memory Bank and Seed Exchange, 135
Landry, Pat, 49

leases, 9, 79, 83–90, 105, 108, 118, 132, 137, 169
Lent, 21, 119, 176n4
lettuce, 28, 30, 115
livelihoods, 9, 46, 54, 154, 156, 160; mixed economies, 175n4; occupational multiplicity, 157; oil and gas work, 9, 37, 49–50, 56, 160, 161. *See also* assemblages
"living off the land," 25, 56, 60, 64, 137, 150, 151
LSU (Louisiana State University), ix, 9, 13, 17, 18, 140, 165, 166
Luton, Harry, viii, 162, 165

maroons, 41, 43
masculinity, 150–51, 169
McIlhenny, A. E., 87
Metzger, Haley, 72–73, 84, 126
middens, 39
mirliton, 31, 122, 177n8
mounds: ceremonial, 39; shell middens, 39
Morgan City, Louisiana, 27, 33, 64, 179n2
mutual aid, 135, 137, 145, 161, 164, 169, 181n1

Natchez Trace, 4
Native Americans, 7, 17, 38–43, 44, 45, 47, 49, 50, 51–55, 58, 59, 60, 114, 141, 177n4, 178n2. *See also* Indigenous people
Neighborhood Story Project, 135
New Orleans, Louisiana, viii, ix, 13, 17, 18, 33, 36, 40, 41, 43–45, 47, 48, 52, 55, 58, 60, 120, 135, 166, 177n8
Nicholls State University, ix, 12, 13, 15, 17, 18, 24, 67, 79, 90, 92, 165; students, 67, 75, 112, 117, 130, 178n8

oil: industry, 9, 36, 37, 49, 50, 56, 160–61; spill, 12, 147, 160. *See also* livelihoods
oysters, 40, 47, 50, 53, 54, 158; oyster middens, 40; oyster spaghetti, vii, viii, x, xi, 4, 68, 157, 167; oystering, 37; oystermen, 49

Parfait, Jessi, 165
pecans, 30, 39, 50, 109, 162
peppers, 20, 52, 109, 115, 122, 124, 144, 145
place, 9, 11, 33–37, 39–41, 53–55, 58, 60–61, 75, 83, 92, 94, 104, 129, 132, 135, 136, 139, 141, 151, 169, 178n3; experience of, 162; of refuge, 161; sense of, 10–11, 26–27, 51–61, 157, 159
plantations, 42–45; Ellenders, 45; Eloise, 43; Klondyke, 50; Minors, 45
Poe, Melissa, 10, 101, 165
potlucks, 21
Pottery, Nancy, 154–55
poverty, 17, 36, 38, 59, 115, 137, 177n2. *See also* inequality
Poverty Point, Louisiana, 4, 175n3
precarity, 160, 182n5
prices, 41, 140; of bait, 67; of club membership, 89–90; driven by relationship and networks, 119, 120; of fur, 47, 49; market price and subsistence use, 101–2; multiple prices (dock, factory, neighbors), 120; overpriced, xi, 116; of shrimp, 120, 140, 160
Probyn, Elspeth, 151, 171
profit maximization, 101–2, 156
Puerto Rico, 156–57
Putnam, Robert, 104

rabbits, 11, 15, 30, 48, 54, 72–74, 107, 122, 125, 145, 147, 148
race, 56–60; hierarchies, 168; identity, 60; labels, 59; masculinity and, 150; multiracial, 60, 181n1; minorities, 56; violence, 57. *See also* ethnicity
real economy, 3, 157. *See also* community economy
reciprocity, 4, 118, 138. *See also* gifts
recreation: commercial, 68; hunting as, 46; as inseparable, 27, 101, 107; as intangible benefit, 104; as problematic category, 138, 151; recreational limit, 20, 105; sports teams, 169; and subsistence practice, 5, 137, 157; versus subsistence, 10
recycling and reuse, 148–49
Red River, 40
Reese, Ashanté, 148
refuge, 36, 41, 54, 55, 161
research methods, vii–ix, xi, 11–16, 153, 167; diversity of researchers, 16–18; drop-in interviews, 15, 67; focus groups, ix, 15, 145, 160; food logs, viii, ix, 18, 19,

68, 108–15, 116–17, 128, 143, 160, 167; freezer inventory, 15, 171; interviews and conversations, 11, 15, 18, 21, 24, 55, 59, 64, 67, 68, 75, 95, 102, 127, 136, 150, 171, 180n6; narrative analysis, 10, 24, 44, 113, 162, 168; oral history, 18, 27, 50, 58, 59, 68, 85, 99, 118, 171, 181n1; participant observation, ix, 15; recruiting students, 15; recruitment flyer, 13–14; tabling at festivals, 13, 15; team meetings, vii, ix, 18, 24; teamwork ethnography, 12–13, 16–18; windshield surveys, 15, 171
Return to Yakni Chitto, 33, 36
river delta geology, 38–39
roadside stands, 121–24

Salmón, Enrique, 5, 138
sassafras, 44, 53, 54, 122, 139, 159, 181n8. *See also* filé
satsumas, vii, xi, 109, 110, 113, 114, 119, 122, 124, 158, 159, 175n2, 178n6, 181n8
Saunders, Mike, 18, 72, 79, 120–21, 125, 152, 165
self-making, 11
self-reliance, 5, 7, 9, 75, 76, 83, 115, 116, 137, 139, 145–51, 154, 159, 162
Serigny, John, 69–71, 85, 92, 94, 95, 126, 136–37, 179n8
sharing, vii, xi, 5, 9, 21, 56, 58, 59, 75, 94, 106, 113–14, 116–19, 137, 138, 143–46, 154, 157, 159, 161, 164, 170, 173, 180n6, 181n1
Smithridge, Louisiana, 58, 60
social capital, 104
social inequality, 38, 116, 148, 177n2
social networks, xi, 4, 9, 27, 94, 102, 117, 119, 120, 124, 161
solidarity, 7, 9
sovereignty, 9, 155. *See also* food: sovereignty
shrimp: barter, 96, 120, 121, 126; boat, 56, 99, 105, 107, 148, 150; boiled, vii, x; boils, 11; deveining, 125; drying platforms, 50, 52, 176n3; factory, 92; family feasts, 67, 75, 106, 117, 133; fishery, 37; frequency, 109; fresh-caught, 10, 31; history, 47, 49; licenses, 51; limits, 136; prices, 120, 160;

role in diet, 141; seasons, 108, 176n1, 181n3; sharing, 110, 114, 116, 117, 118, 119, 143, 158
Shrimp and Petroleum Festival, 179n2
shrimping, ix, 19, 25, 139, 140, 142, 178n5; and autonomy, 148; commercial, 21–22, 37, 127; full-time, 20; and gender identity, 169; as heritage, 162; learning to shrimp, 76, 130, 150
Sirois, Connie, 24
Sothern, Tyler, 117
Sourdelier, Jerome, 19–21, 22, 24, 105–7
Spanish people, 42, 45, 60, 177n5. *See also* Isleño people
St. Malo, Louisiana, 177n7
sugar, 37, 42, 44, 45, 48, 50, 52, 57–58, 177n6
subsistence: in Alaska, 5–6, 173; in Canada, 132; coexistence with oil, 50; commonplace, 111; conventional meanings, 3–4; cultural revitalization; 5, 36, 55; defined, viii, xi, 3–11, 23–24, 25, 31, 101–2, 153–54, 155; diverse identities, 53; in film, 65, 170; frequency, 109; futures, 159–62; heritage, 61; heterogenous, 132; and Indigenous people, 53–55; investments, 125–31; land claim, 138; long-distance trade, 4; many names, 3; narrative, 10; on the Potomac and Anacostia Rivers, 6, 58; as practice, xi, 5, 7, 9, 145; in Puerto Rico, 157; and raising children, 10, 65, 72–77, 154, 158; resilience, 161; saturation (density), 122; snapshots, 99–108; stereotypes, 83, 136; table of word associations, 137; values, x, 6, 9, 60, 65, 75–76, 96–97, 101, 106, 120, 137, 139, 141–42, 145, 150, 162, 181n7; versus recreation, 155–57
subsistence practices, viii, xi, xii, 3, 5–11, 15, 19–31, 38, 52, 58, 61, 65, 76, 82, 101–2, 115, 116, 122, 131, 132, 136–40, 145, 148, 152–55, 157–62, 168, 173, 175n4, 180n1 (chap. 7), 182n5; "communities of practice," 10
squirrels, 30, 128
syndemics, 177n2
Swamp People, 82, 83, 136

taste, x–xi, 26, 150, 167; delicious, 66, 76; and hard work, 131; pleasure, 159; pride, 150; of reality, 141; and valued foods, 171
TaWaSi social club, 22, 24
Terrebonne Parish, ix, 12, 18, 24, 30, 33–34, 37, 38–49, 53, 57–58, 67, 93, 97, 122–23, 129, 136, 161, 164, 175n3, 177nn6–7, 178n2, 180n5
Texaco, 49
Thibodaux, Louisiana, vii, 16, 28, 33, 45, 48, 57, 64, 79, 99, 120, 123, 179n2, 180n1; massacre, 57–58
Tidwell, Mike, 56
totalizing, 6–7, 176n8
Trahan, Glynn, 21–24, 27, 64, 118, 139, 143, 148, 158
trapping, 41, 44, 46–49, 50, 64, 70, 79, 160; trappers war, 48
Tsing, Anna, 7, 9, 18, 160

Underhill, Ruth, 47
United Houma Nation. *See* Houma (United Houma Nation)

value, 5, 6, 10, 99–134, 169, 181n7; of autonomy, 16; of camp, 70, 89; as part of the economy, 3; of sharing, 143–44; of specific foods, 9, 168, 171; story, x; and youth, 67–68
values, 56, 60, 137, 139, 162; documentation and awareness, 157; enculturation, 65, 75–76, 96–97, 158; self-reliance and autonomy, 145–51; transmission, 141
vegetables, harvesting frequency, 109, 111, 113. *See also* sharing
Verdin, Monique, 33, 36, 54, 60, 135–36, 161
Voisin, Magnus, 50

wage: economy, 9; of independent fishers, 38, 140; work, 4, 10, 50, 104, 117, 120, 137, 150, 157
Wali, Alaka, 156, 165
well-being, 5, 7, 9, 10, 65–66, 77, 102, 105, 115, 145, 154, 155–57, 161–62, 169, 177n2
white beans, 21, 52, 114

white people, 7, 38, 53, 55, 56, 57, 59, 87; Black-white divide, 60; militias, 57; settlers, 168; working-class men, 182n4
wild boar, ix, 46
wildflowers, 28, 30
Wildlife & Fisheries, 16, 46, 85, 127, 136, 155, 168
Wilmore, Cherry, 58

ABOUT THE AUTHORS

Helen A. Regis is a cultural anthropologist in the Department of Geography and Anthropology, Louisiana State University. She has been involved in urban, public, collaborative, and applied anthropology for over two decades, working in the southern US, Europe, and West Africa. Her work engages the intersections of culture and commerce, the grassroots and public policy with a focus on public space, social movements, and cultural heritage, coastal cities, and sustainable tourism. Regis is series editor and board president of the Neighborhood Story Project, an organization that creates collaborative ethnographies, exhibits, and events in New Orleans. Publications include "Ships on the Wall: Retracing African Trade Routes from Marseille, France," *Genealogy* (2021); "Putting the Ninth Ward on the Map: Race, Place, and Transformation in Desire, New Orleans," (with Rachel Breunlin, 2006) *American Anthropologist*; *Charitable Choices: Religion, Race, and Poverty in the Post-Welfare Era* (with John Bartkowski, 2004).

Shana Walton teaches in the Department of English, Modern Languages and Cultural Studies at Nicholls State University in Thibodaux, Louisiana, and she is on the advisory committee of the university's Center for Bayou Studies. She is a founding member of the Bayou Culture Collaborative, a project designed to connect those interested in the impact of Louisiana's land loss on the culture and heritage of south Louisiana. She coedited (with Nathalie Dajko) *Language in Louisiana: Community and Culture* (2019), and (with Barbara Carpenter) *Ethnic Heritage in Mississippi: The Twentieth Century*, (2012).